New Frontiers of Network Analysis in Systems Biology

Avi Ma'ayan • Ben D. MacArthur
Editors

New Frontiers of Network Analysis in Systems Biology

Springer

Editors
Avi Ma'ayan
Pharmacology and Systems Therapeutics
Mount Sinai School of Medicine
NY, USA

Ben D. MacArthur
Faculty of Medicine and School
of Mathematics
Institute for Life Sciences
University of Southampton
Southampton, UK

ISBN 978-94-007-4329-8 ISBN 978-94-007-4330-4 (eBook)
DOI 10.1007/978-94-007-4330-4
Springer Dordrecht Heidelberg New York London

Library of Congress Control Number: 2012942164

Printed on acid-free paper

Springer is part of Springer Science+Business Media (www.springer.com)

Preface

As progress in the biomedical sciences leads to breakthroughs in biotechnology, biotechnology is, in turn, fueling progress in the biomedical sciences. Biotechnological advances in instruments capable of measuring molecular components within cells at a genome-wide level, as well as the infusion of ideas from physics and mathematics for data analysis and advances from computer science for data-sharing, storage, search and visualization, are all expected to collectively lead to a surge in biomedical breakthroughs in translational research in the near future. Systems biology is playing an increasingly important role in many aspects of biomedical research [1]. Emerging as the result of the sequencing of yeast, mouse and human genomes, systems biology aims to develop robust theoretical models that explain how molecular components give rise to cellular, tissue and organism phenotypes at the genome-wide scale [2]. In addition, as molecular data collection in mammalian cell biology research becomes more extensive and complex, advances in computation are now critical for continuing progress [3]. Such progress may enable us, amongst other things, to better control cell behavior with small molecules (and perhaps other means) and, in turn, translate these developments to therapeutics. This collection of invited reviews describes some of the latest cutting-edge experimental technologies and their applications in systems biology, as well as new computational strategies to model the data generated from these new experimental techniques. In particular, this collection focuses on the study of mammalian embryonic stem cells, technologies involving mass-spectrometry proteomics and single cell measurements, methods for modeling complex stochastic systems, network-based classification algorithms and the emerging field of systems pharmacology.

The importance of cell-to-cell variability in a given tissue or in cell culture and in particular the presence of a stem cell subpopulation in tissues or tumors is gaining awareness in the systems biology community. Methods to measure the abundance of molecular species in single cells are critical for understanding the mechanisms underlying tissue heterogeneity and the dynamics of cell-to-cell communication. In Chap. 1, Fumio Arai et al. describe cutting-edge technologies used to track gene expression in single cells. Individual cells need to constantly make cell-fate decisions. For instance, stem cells must decide whether to remain stem-like or differentiate. Such decisions are made in response to intrinsic and extrinsic cues that signal to gene regulatory networks, typically containing positive and negative feedback loops. In Chap. 2, Sonya Ridden and Ben D. MacArthur describe several mathematical modeling approaches that can be used to help us to understand how such regulatory circuits function dynamically. In addition to deterministic control by regulatory networks, cell fate decisions are also made with the assistance of molecular fluctuations or "noise". Therefore, in Chap. 3 Srinandan Dasmahapatra discusses different ways in which molecular noise can affect cell fate at both the cellular and population levels. In Chap. 4, Thilo Gross introduces a novel rational approach, known as generalized models, for modeling the dynamics of complex systems in the absence of detailed regulatory information. He then provides some examples on how these rigorous principles can be applied to problems in systems biology.

In order to simulate the dynamics of regulatory networks, measurements of protein abundance and protein–protein interactions are necessary to inform and constrain dynamical models. In addition, in the field of developmental biology the study of mammalian embryonic and induced pluripotent stem cells has seen a recent surge in interest. In particular, many high-throughput technologies are now being used to quantify regulatory components and interactions in mammalian stem cell biology. In Chap. 5, Richard Unwin et al. describe the important technique of mass-spectrometry isobaric tagging for relative and absolute protein quantification, illustrating how it may be applied to profile the proteome of mouse embryonic and hematopoietic stem cells. In Chap. 6, Jianlong Wang brings together stem cell research and proteomics by describing how mass-spectrometry proteomics may be used to unravel the intricate protein–protein interaction networks governing self-renewal and pluripotency of embryonic stem cells. In a complementary chapter (Chap. 7), Ana Sevilla et al. discuss emerging datasets which are being collected to profile embryonic stem cells at different "layers" of molecular regulation (genetic, proteomic, epigenetic, etc.). They then survey different tools and methods for collecting and analyzing these complex datasets and discuss rational approaches for integrating knowledge across regulatory layers. In Chap. 8, Anna Malovannaya et al., who recently published a massive immuno-precipitation followed by mass-spectrometry (IP-MS) study in the journal *Cell* [4], describe the experimental and computational challenges and opportunities faced when using mass-spectrometry proteomics techniques to better reconstruct the human interactome. Their work is specifically geared toward unraveling the molecular complexes in the nucleus responsible for hormonal regulation of transcription.

In Chap. 9, Ruth Dannenfelser et al. provide a brief introduction to an emerging offshoot from systems biology known as systems pharmacology. They discuss how data integration can be used to build network-based classifiers that can be used to predict new roles for existing and experimental drugs and predict side effects for new drugs based on the information about existing drugs. Finally, on a related topic, in Chap. 10 Bernhard Schuldt et al. ask "What can networks do for you?" and discuss their experience in developing network-based classification algorithms in the stem cell field as well as the uses and abuses of networks in systems biology more generally.

In conclusion, this book aims to provide a wide range of overview articles on some of the most exciting emerging new directions taken by the expanding field of systems biology. It contains an overview of cutting-edge computational and experimental methods and their applications in mammalian regulatory biology, with particular focus on stem cells, transcriptional regulation by proteins, proteomics and network-based classification and systems pharmacology.

New York and Southampton

Avi Ma'ayan
Ben D. MacArthur

References

1. Ma'ayan A (2008) Network integration and graph analysis in mammalian molecular systems biology. IET Syst Biol 2(5):206–221
2. Ma'ayan A, Blitzer RD, Iyengar R (2005) Toward predictive models of mammalian cells. Annu Rev Biophys Biomol Struct 34(1):319–349
3. Ma'ayan A (2009) Insights into the organization of biochemical regulatory networks using graph theory analyses. J Biol Chem 284(9):5451–5455
4. Malovannaya A, Lanz Rainer B, Jung Sung Y, Bulynko Y, Le Nguyen T, Chan Doug W, Ding C, Shi Y, Yucer N, Krenciute G et al (2011) Analysis of the human endogenous coregulator complexome. Cell 145(5):787–799

Contents

Chapter 1
Gene Expression Profiling and Regulatory Networks in Single Cells

Fumio Arai, Kentaro Hosokawa, Yoshiko Matsumoto, Hirofumi Toyama, and Toshio Suda

Abstract Tissues and organs are composed of various kinds of cells including stem cells, progenitors, and terminally differentiated cells that have large variations of gene expression. Within a seemingly homogeneous cell population, gene expression levels may differ dramatically on a cell-to-cell level. However, differences or variations of individual cells are masked by the averaging effect of pooled samples in population analysis. Therefore, analyzing transcript levels for multiple genes across multiple individual cells could be key for understanding the unique characteristics of individual cells and for clarifying the complicated mechanisms controlling the function of individual cells. Recent advances in cDNA amplification techniques from single cell samples allow for analyzing gene expression patterns in individual cells. Amplified cDNAs are applicable to measurements of gene expression systems such as micro-arrays, DNA sequencing, and TaqMan assay-based high throughput nanofluidic real time PCR array analysis. Gene expression profiling at the single cell level allows us to identify specific sub-populations in heterogeneous cell populations in tissues and clarify the complex networks controlling the function of individual cells.

Keywords Single cell • cDNA amplification • Nanofluidic real time PCR array

1.1 Introduction

Tissues and organs consist of a multiplicity of cells that are highly variable in their function and phenotype. This variety of cells may be due to differential gene expression, and quantitative determination of gene expression in individual cells

F. Arai (✉) • K. Hosokawa • Y. Matsumoto • H. Toyama • T. Suda
Department of Cell Differentiation, The Sakaguchi Laboratory of Developmental Biology, School of Medicine, Keio University, 35 Shinano-machi, Shinjuku-ku, Tokyo 160-8582, Japan
e-mail: farai@a3.keio.jp

A. Ma'ayan and Ben D. MacArthur (eds.), *New Frontiers of Network Analysis in Systems Biology*, DOI 10.1007/978-94-007-4330-4_1,

provides valuable information for understanding the molecular mechanisms that control cell behavior in tissues, especially stem cells.

Gene expression profiling is an essential tool for many biological investigations. For example, microarrays are powerful tools for such studies. Conventional gene expression analyses usually require large amounts of starting material obtained from cell populations or target tissues/organs. However, rare populations in the tissue, in particular stem cells, are not always easily isolated in such quantities. The solution is to examine multiple numbers of individual cells to identify the variation of individual characteristics.

Recent advances and refinement of techniques for single cell cDNA amplification, quantitative real-time polymerase chain reaction (PCR), microarray analysis, and DNA sequencing allow for analyzing of gene expression patterns in single cells. By using gene expression profiling at the single cell level we can identify genes expressed in particular cell populations, such as tissue stem cells. In addition, we can identify specific sub-populations in heterogeneous cell populations in tissues, based on gene expression analysis at the single cell level. Single cell gene expression analysis can reveal differences among cells, even in cell populations that were thought to be homogeneous. Moreover, the ability to profile individual cells is required to discover how different RNA/protein expression levels in the cell naturally lead to different regulatory networks being active/inactive in different cells within the same population. By looking at each individual cell instead of at the population average, correlations in expression between genes can be identified and regulatory networks inferred.

Although single cell gene expression analysis technology is new, single cell studies of stem cells allow us to address the behavior of single stem cells within a complex population of cells. Analyzing transcript levels for multiple genes across multiple single cell samples will therefore be important for understanding key biological processes such as stem cell development, tumoriogenesis, and disease progression.

1.2 The Significance of Single Cell Analysis

Generally, measurements of gene expression are taken on a large number of cells. However, studies with samples comprised of pooled cells, generate relative expression levels that represent an average of all cells assayed. Furthermore, tissues are composed of different types of cells that have large variations in gene expression. In addition, within highly purified cell populations, such as tissue stem cells, that are supposedly homogeneous, it is possible that gene expression patterns differ dramatically on a cell-to-cell level even after careful cell sorting. Individual stem cells can have different cell division rates and have different self-renewal/differentiation potentials. Indeed, the highly purified long-term hematopoietic stem cell (LT-HSC) population comprises two subsets: a dormant population (about 15 %) dividing only about every 145 days, and an 'active' population (about 85 %) dividing about once every month, though these two populations present the same surface phenotypes, $Lin^{-}Sca\text{-}1^{+}c\text{-}Kit^{+}$ (LSK) $CD34^{-}CD48^{-}CD150^{+}$ [43]. These data indicate

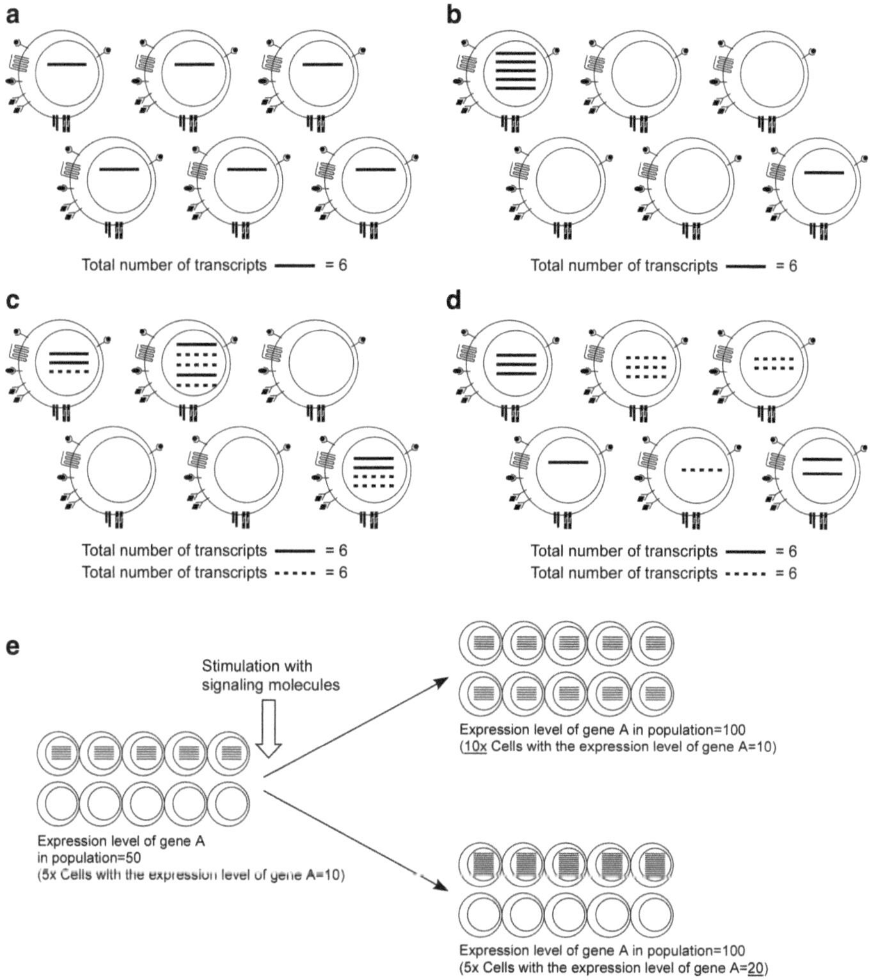

Fig. 1.1 Single cell analysis is required for the identification of the variation of individual cell characteristics. Within a seemingly homogeneous cell population, variations in gene expression may differ dramatically on a cell-to-cell level. Measuring gene expression at the population levels obscures these differences. Single cell gene expression analysis can reveal differences among cells even in populations that were purified by the same cell surface phenotypes. Single cell measurements can distinguish the differences between (**a**) and (**b**), and between (**c**) and (**d**). Panel (**e**) shows some possibilities for the increase of the expression level of "A" in the population after stimulation with signaling molecules

that the highly purified LT-HSC population contains more specific sub-populations in the bone marrow (BM).

Ståhlberg and Bengtsson [36] review two key points of concern when profiling heterogeneous populations (see Fig. 1.1). (1) Population analyses will not reveal how a particular transcript is distributed among the cells. Analysis of gene expression in pooled samples easily masks differences in individual cell characteristics.

Fig. 1.2 Single cell profiling is useful for the identification of single cell genetic regulatory networks. A simple hypothetical transcriptional regulatory network consisting of three interconnected "modules". Each node is a transcription factor and edges denote transcriptional regulation of gene expression. Depending on which transcription factors are present in the cell different modules will be active, and the others inactive. Active modules are in *black* and inactive modules are in *grey*: (*left*) only the module on the *top right* is active; (*middle*) the *top* two interconnected modules are active; (*right*) a feedback loop running through all three modules is active

(2) Gene expression analyses in cell populations lead to the oversight of potentially important gene correlations. Indeed, it is difficult to distinguish whether transcripts "A" and "B" are co-expressed by all cells within the population or are separately expressed in different cell sub-types within the population. In addition to these two points, it is also difficult to distinguish whether changes (in response to external stimulation such as cytokines, for example) in population-level gene expression levels are due to changes in gene expression levels in all cells in the population and/or changes in the number of cells expressing particular genes.

Single cell gene expression profiling is also essential for identifying the genetic regulatory networks that directly regulate the behavior of individual cells within (immunophenotypically homogenous) populations. For example, when different transcripts are present different parts of the underlying regulatory network become "active" and different parts "inactive" enabling the cells to behave in different ways (Fig. 1.2).

Therefore, the examination of multiple numbers of individual cells is required for the identification of the unique mix of individual cells in the population and for clarifying the molecular mechanisms controlling the function and fate of individual cells.

1.3 Methods for Gene Expression Analysis at the Single Cell Level

DNA microarray, real-time PCR, and high throughput sequencing analyses are used for detection of gene expression. It is known that a typical single cell contains ~1 pg mRNA, corresponding to a few hundred thousand molecules transcribed from about ten thousand genes [17]. Therefore, methods for single cell measurement generally require amplification of cDNA. Exponential amplification (based on PCR) [3, 4] and multiple-round linear amplification (based on in vitro transcription with T7

RNA polymerase) [6, 41] of mRNAs are two methods which have been developed for measuring gene expression of cell populations. For example, Kurimoto et al. developed a protocol for cDNA amplification from single cell samples using a small number of directional PCR cycles followed by a round of linear amplification that provides a template for high-density oligonucleotide microarray analysis [20, 21]. Similarly, Esumi et al. [8] also reported a method for single cell microarray analysis. In their protocol, cDNA from single cell samples was amplified with a combination of Super SMART PCR amplification and T7 RNA polymerase amplification. Reverse transcription (RT) and specific target amplification (STA) using TaqMan Gene Expression Assay Mixes (Applied Biosystems) are also useful for real time PCR-based gene expression analysis at the single cell level. STA allows for a multiplexed preamplification of up to 100 genes by using a 0.2x pool of gene expression assays as the source of primers. By using the same assays in the preamplification reaction as the real time PCR reaction, only the targets of interest are amplified.

1.4 High Throughput Real Time PCR Array Analysis

DNA microarrays are extremely powerful tools for biological and medical studies since they allow one to probe virtually the entire transcriptome and give an overall image of gene expression. These data usually need to be validated with real time PCR experiments, which has high sensitivity, dynamic range, and reproducibility. However, conventional real time quantitative PCR is a low-throughput technique, which limits the number genes that can be verified in a single experiment. Furthermore, many microarray studies are concluded by identifying a set of 20–100 genes that are the most important determinants of the biological phenomenon of interest [34]. Validation studies or practical medical applications then require measurement of those genes on much larger sample sets than are practical with conventional microarrays.

Microfluidic technology has been applied to a number of laboratory techniques including gene expression analysis [35]. Quake and his colleagues produced microfluidic devices that allow control and manipulation of very small volumes of liquid [40] and microfluidic chip-based devises can now be used for single cell measurement of gene expression. For example, using a microfluidic chip-based real time PCR array (a product of the Fluidigm Corporation), Warren et al. [42] measured the expression of transcription factors that play a critical role in the lineage commitment of hematopoietic progenitors at the single cell level. For the measurement of a larger number of target genes in multiple single cell samples, the Dynamic Array (another product of the Fluidigm Corporation) can be used with real time PCR. Single Dynamic Array chips may be used to measure the expression of 48 or 96 genes in 48 or 96 individual different samples, respectively. In combination with the STA technique, the Dynamic Array allows gene expression analysis of up to 96 single cell samples against 96 genes, giving 9126 individual real time PCR reactions.

Such single cell real time PCR arrays are useful for the dissection of gene regulatory networks. For instance, Foygel et al. [10] performed microarray experiments on

Fig. 1.3 Single cell real time PCR array analysis reveals the heterogeneity of LT-HSCs. Heat maps of the single-cell real time PCR array analysis of selected genes in LSK-CD41-CD48-CD150 + CD34- LT-HSC cells. The horizontal axis in the heat map is a list of genes tested, including internal controls, and the vertical axis represents cell samples. The color in each cross point indicates threshold cycles. Gene expression in highly purified LT-HSCs was highly heterogeneous

pooled early embryos (two cell mouse embryos) in which *Oct3/4* (POU domain, class 5, transcription factor 1 (*Pou5f1*)), an essential regulator for pluripotency of embryonic stem cells, was knocked down and validated the expression of 42 genes at the single cell level. They found that *Oct3/4* plays a critical role in embryonic development prior to the blastcyst stage. Interestingly, *Rest* and *Mta2*, both of which play critical functions in regulating pluripotency of ES cells, were found be the most tightly regulated by *Oct3/4*.

Single cell real time PCR arrays are also suitable for the investigation of the variation in individual cells in other stem cell fractions. It has been reported that the highly purified long-term hematopoietic stem cells (LT-HSCs) are a heterogeneous population in their cell cycle (dormant and active states) and self-renewal capacity [2, 43]. Although it is still unclear whether variation of gene expression in individual LT-HSCs is responsible for the control of the cell cycle and self-renewal capacity of LT-HSCs, we have found variation in gene expression in the LT-HSC fraction at the single cell level (see Fig. 1.3).

Since microarray and real time PCR array analysis have advantages and disadvantages, it is important to select the experimental platform in accordance with the aim of the study. Traditional microarrays or next generation sequencing platforms may be used initially to perform broad gene discovery research. On the other hand, real time PCR array systems are ideally suited for focused gene expression analysis and hypothesis testing.

1.5 Identification of Subpopulation of Bone Marrow Stromal Cells Using Single Cell Gene Expression Analysis: A Case Study

The interaction of stem cells with their supportive microenvironmental niche, which is composed of cellular components located around stem cells, facilitates the signaling networks that control the balance between self-renewal and differentiation [26, 27, 44]. Therefore, identification of the molecular interaction of tissue stem cells with niches is critical for understanding the regulatory mechanisms of stem cells. In the hematopoietic system, the bone marrow (BM) supports both self-renewal and differentiation of hematopoietic stem cells (HSCs) in specialized niches, endosteum (the border between the bone and the BM) and in perivascular sites. Cytokines, chemokines, adhesion molecules, proteolytic enzymes, neurotransmitters and transcription factors regulate the balance between quiescence and activation (proliferation and migration) of HSCs [22, 44]. In particular, Angiopoietin-1 (Angpt1) [1], Kit-ligand (Kitl)/Stem cell factor (SCF) [13, 39], chemokine (C-X-C motif) ligand 12 (Cxcl12) [16, 19, 24, 38], Thrombopoietin (THPO) [31, 46], Wnt [9], Jagged1 (Jag1) [5], osteopontin (OPN) [30, 37], and N-cadherin (Cdh2) [12, 14, 15] are known to be involved in the niche regulation of HSCs in BM, and cooperative regulation among cytokine signals and cell adhesion molecules is required for the maintenance of HSCs. In the BM niche, many types of cells participate the regulation and maintenance of LT-HSCs. So far, it has been reported that osteoblasts, other mesenchymal cells including mesenchymal stem/progenitor cells (MSCs/MPCs), reticular cells, adipocytes, osteoclasts, other stromal cells, and regulatory T cell have also been implicated in regulation of HSC maintenance in the niche [11, 25, 29, 33, 38]. Furthermore, the cells in the endosteum are a heterogeneous population in terms of their degree of differentiation and accompanying functions [18, 45]. The precise cellular and molecular contribution of endosteal cell populations to the HSC-supportive microenvironment is still unclear.

Recently, we identified subpopulations of adult BM endosteal cells based on the expression of ALCAM and Sca-1. Non-hematopoietic and non-endothelial cells were enriched in the $CD45^{-}CD31^{-}Ter119^{-}$ fraction and further subdivided into three fractions: $ALCAM^{+}Sca\text{-}1^{-}$, $ALCAM^{-}Sca\text{-}1^{+}$, and $ALCAM^{-}Sca\text{-}1^{-}$ cells [28]. Analysis of differentiation potential and conventional gene expression PCR analysis revealed that $ALCAM^{+}Sca\text{-}1^{-}$, $ALCAM^{-}Sca\text{-}1^{-}$, and $ALCAM^{-}Sca\text{-}1^{+}$ cells are mature osteoblasts, immature osteoblasts, and mesenchymal progenitor cells, respectively.

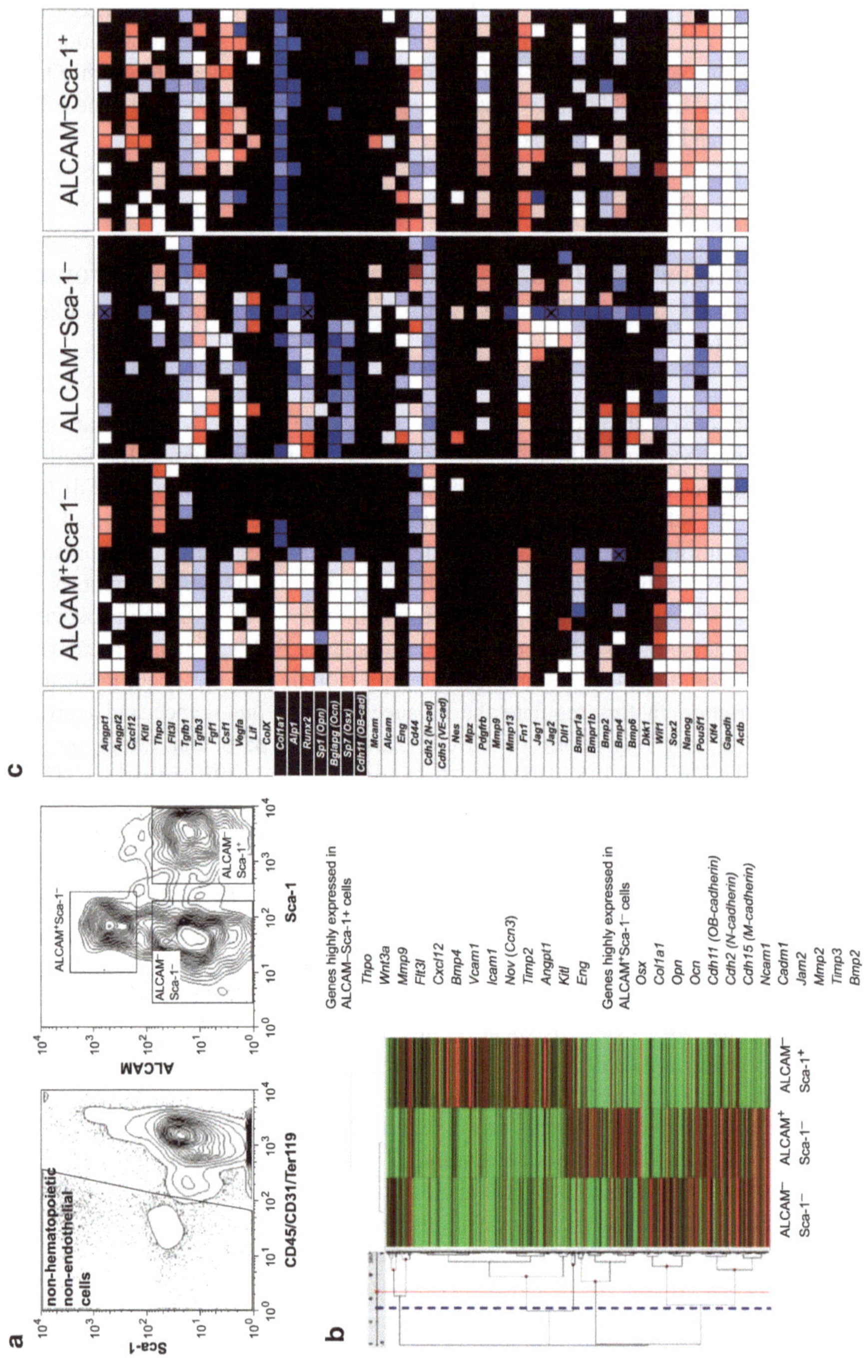
a
non-hematopoietic
non-endothelial
cells
Sca-1
CD45/CD31/Ter119
ALCAM
Sca-1
ALCAM+Sca-1−
ALCAM−
Sca-1−
ALCAM−
Sca-1+
b
ALCAM−
Sca-1−
ALCAM+
Sca-1−
ALCAM−
Sca-1+
Genes highly expressed in
ALCAM−Sca-1+ cells
Thpo
Wnt3a
Mmp9
Flt3l
Cxcl12
Bmp4
Vcam1
Icam1
Nov (Ccn3)
Timp2
Angpt1
Kitl
Eng
Genes highly expressed in
ALCAM+Sca-1− cells
Osx
Col1a1
Opn
Ocn
Cdh11 (OB-cadherin)
Cdh2 (N-cadherin)
Cdh15 (M-cadherin)
Ncam1
Cadm1
Jam2
Mmp2
Timp3
Bmp2
c
ALCAM+Sca-1−
ALCAM−Sca-1−
ALCAM−Sca-1+
Angpt1
Angpt2
Cxcl12
Kitl
Thpo
Flt3l
Tgfb1
Tgfb3
Fgf1
Csf1
Vegfa
Lif
ColX
Col1a1
Alp1
Runx2
Sp1 (Opn)
Bglapg (Ocn)
Sp7 (Osx)
Cdh11 (OB-cad)
Mcam
Alcam
Eng
Cd44
Cdh2 (N-cad)
Cdh5 (VE-cad)
Nes
Mpz
Pdgfrb
Mmp9
Mmp13
Fn1
Jag1
Jag2
Dll1
Bmpr1a
Bmpr1b
Bmp2
Bmp4
Bmp6
Dkk1
Wif1
Sox2
Nanog
Pou5f1
Klf4
Gapdh
Actb

Fig. 1.4 Identification of an immature population in bone-lining osteoblasts using single cell gene expression analysis. (**a**) Representative FACS profile of ALCAM and Sca-1 expression in CD45$^-$CD31$^-$Ter119$^-$ cells in the bone-associated fraction. Non-hematopoietic and non-endothelial cells were enriched in CD45$^-$CD31$^-$Ter119$^-$ cells, and were gated and the expression of Sca-1 and ALCAM was analyzed. CD45$^-$CD31$^-$Ter119$^-$ cells were subdivided into three fractions: ALCAM$^+$Sca-1$^-$, ALCAM$^-$Sca-1$^-$, and ALCAM$^-$Sca-1$^+$ cells. (**b**) Conventional microarray analysis of three endosteal populations. ALCAM$^+$Sca-1$^-$ cells highly expressed osteoblastic marker genes such as *Osx*, *Col1a1*, *Opn*, *Ocn*, and *Cdh11* (*OB-cadherin*). In contrast, ALCAM$^-$Sca-1$^+$ cells highly expressed the mesenchymal stem/progenitor cell marker, *Eng*. The ALCAM$^+$Sca-1$^-$ fraction expressed multiple cell adhesion molecules, such as *Cdh11*, *Cdh2* (*N-cadherin*), *Ncam1*, *Cadm1*, and *Jam2*, whereas the ALCAM$^-$Sca-1$^+$ fraction highly expressed cytokine-related genes, such as *Thpo*, *Wnt3a*, *Cxcl12*, and *Angpt1*, at a higher level than the other fractions. (**c**) Representative heat map of the single-cell real time PCR array analysis of selected genes in ALCAM$^+$Sca-1$^-$, ALCAM$^-$Sca-1$^-$, and ALCAM$^-$Sca-1$^+$ cells. The *vertical axis* in the heat map is a list of genes tested, including internal controls, and the horizontal axis represents cell samples (16 independent cells in each fraction/experiment). Colors indicate relative expression values

Among these three endosteal cell populations, ALCAM$^+$Sca-1$^-$ cells showed robust supporting activity for HSCs in *in vitro* culture. Microarray analysis showed that cytokine-related genes were highly expressed in the ALCAM$^-$Sca-1$^+$ fraction. In contrast, compared to other fractions, the ALCAM$^+$Sca-1$^-$ fraction showed higher levels of expression of genes related to cell adhesion (Fig. 1.4). In order to illustrate the use of single cell profiling, we now present a case study in which heterogeneity in these bone marrow stromal cell populations is explored using single cell profiling.

By using single-cell real time PCR array analysis, we further characterized the three endosteal cell populations described above (Fig. 1.4). We analyzed the expression of 39 genes including osteoblastic markers, cytokines, cell adhesion molecules, and stem cell markers in single endosteal cell samples of the three endosteal fractions. As a general trend, gene expression patterns in each population are consistent with the results of the bulk measurement using conventional real time PCR and microarray. Interestingly, using single cell gene expression analysis, we found ALCAM$^+$Sca-1$^-$ cells are a heterogeneous population and identified subpopulation of ALCAM$^+$Sca-1$^-$ cells that showed very low or no expression of osteoblastic markers (about 36 % of ALCAM$^+$Sca-1$^-$ cells). Furthermore, approximately 42.3 % of osteoblastic marker$^{low/-}$ ALCAM$^+$Sca-1$^-$ cells expressed at least one BM HSC niche-related cytokine gene. More interestingly, about 30 % of osteoblastic marker$^{low/-}$ cells expressed at least one pluripotent stem cell marker such as *Sox2*, *Oct3/4*, or Nanog homeobox (*Nanog*) at relatively high levels compared with other ALCAM$^+$Sca-1$^-$ or ALCAM$^-$Sca-1$^+$ cells. These data indicate that some of the pluripotent stem cell marker expressing osteoblastic marker$^{low/-}$ ALCAM$^+$Sca-1$^-$ cells might be an immature niche cell population.

It would be interesting to elucidate the potential of the osteoblastic marker$^{low/-}$ ALCAM$^+$Sca-1$^-$ cells, in terms of their self-renewal activity, differentiation potential, and their function in the maintenance of LT-HSCs. In addition, more detailed fractionation of endosteal cells based on single-cell gene expression may provide detailed characterization of niche cell components.

1.6 Conclusion

Single cell gene expression profiling is a powerful tool to shed light on the complex phenomena regulating cell behavior. For instance, since stem cells are not necessarily always in the same state, each stem cell needs to be analyzed as an individual in order to understand the molecular mechanisms of stem cell fate decisions. Analyzing transcript levels for multiple genes across multiple individual cells could be key for understanding the earliest steps in the differentiation of stem cells. On the other hand, it has been shown that gene expression in stem cells does not necessarily correspond to protein expression [23], indicating that transcriptional and post-transcriptional regulation play a critical part in stem cell fate decisions. However, due to technical limitations, it is currently extremely difficult to carry out proteomic studies at the single cell level. Although the numbers of analyzable proteins are limited, the continuous single cell imaging reported by Schroeder and colleagues is useful for lineage tracing of stem cells [7, 32]. Combinational use of multiple techniques will be crucial in order to better understand the complex regulatory mechanisms controlling stem cell fate decisions.

1.7 Methods

For the single cell gene expression analysis shown in Figs. 1.3 and 1.4, we performed RT, STA, and real time PCR arrays. In brief, target cells were sorted directly into the mixture of RT and STA buffer/enzyme mixture, and RT and STA were serially performed. After that, gene expression was analyzed using the BioMark 96·96 (Fig. 1.3) or 48·48 (Fig. 1.4) Dynamic Array (Fluidigm) described above.

Acknowledgments The Funding Program for Next Generation World-Leading Researchers (NEXT Program) supported this work.

References

1. Arai F, Hirao A, Ohmura M, Sato H, Matsuoka S, Takubo K, Ito K, Koh GY, Suda T (2004) Tie2/angiopoietin-1 signaling regulates hematopoietic stem cell quiescence in the bone marrow niche. Cell 118:149–161
2. Benveniste P, Frelin C, Janmohamed S, Barbara M, Herrington R, Hyam D, Iscove NN (2010) Intermediate-term hematopoietic stem cells with extended but time-limited reconstitution potential. Cell Stem Cell 6:48–58
3. Brady G, Barbara M, Iscove NN (1990) Representative in vitro cDNA amplification from individual hemopoietic cells and colonies. Method Mol Cell Biol 2:17–25
4. Brady G, Iscove NN (1993) Construction of cDNA libraries from single cells. Methods Enzymol 225:611–623
5. Calvi LM, Adams GB, Weibrecht KW, Weber JM, Olson DP, Knight MC, Martin RP, Schipani E, Divieti P, Bringhurst FR et al (2003) Osteoblastic cells regulate the haematopoietic stem cell niche. Nature 425:841–846
6. Eberwine J, Yeh H, Miyashiro K, Cao Y, Nair S, Finnell R, Zettel M, Coleman P (1992) Analysis of gene expression in single live neurons. Proc Natl Acad Sci U S A 89:3010–3014
7. Eilken HM, Nishikawa S, Schroeder T (2009) Continuous single-cell imaging of blood generation from haemogenic endothelium. Nature 457:896–900
8. Esumi S, Wu SX, Yanagawa Y, Obata K, Sugimoto Y, Tamamaki N (2008) Method for single-cell microarray analysis and application to gene-expression profiling of GABAergic neuron progenitors. Neurosci Res 60:439–451
9. Fleming HE, Janzen V, Lo Celso C, Guo J, Leahy KM, Kronenberg HM, Scadden DT (2008) Wnt signaling in the niche enforces hematopoietic stem cell quiescence and is necessary to preserve self-renewal in vivo. Cell Stem Cell 2:274–283
10. Foygel K, Choi B, Jun S, Leong DE, Lee A, Wong CC, Zuo E, Eckart M, Reijo Pera RA, Wong WH et al (2008) A novel and critical role for Oct4 as a regulator of the maternal-embryonic transition. PLoS One 3:e4109

11. Fujisaki J, Wu J, Carlson AL, Silberstein L, Putheti P, Larocca R, Gao W, Saito TI, Lo Celso C, Tsuyuzaki H et al (2011) In vivo imaging of Treg cells providing immune privilege to the haematopoietic stem-cell niche. Nature 474:216–219
12. Haug JS, He XC, Grindley JC, Wunderlich JP, Gaudenz K, Ross JT, Paulson A, Wagner KP, Xie Y, Zhu R et al (2008) N-cadherin expression level distinguishes reserved versus primed states of hematopoietic stem cells. Cell Stem Cell 2:367–379
13. Heissig B, Hattori K, Dias S, Friedrich M, Ferris B, Hackett NR, Crystal RG, Besmer P, Lyden D, Moore MA et al (2002) Recruitment of stem and progenitor cells from the bone marrow niche requires MMP-9 mediated release of kit-ligand. Cell 109:625–637
14. Hosokawa K, Arai F, Yoshihara H, Iwasaki H, Hembree M, Yin T, Nakamura Y, Gomei Y, Takubo K, Shima H et al (2010) Cadherin-based adhesion is a potential target for niche manipulation to protect hematopoietic stem cells in adult bone marrow. Cell Stem Cell 6:194–198
15. Hosokawa K, Arai F, Yoshihara H, Nakamura Y, Gomei Y, Iwasaki H, Miyamoto K, Shima H, Ito K, Suda T (2007) Function of oxidative stress in the regulation of hematopoietic stem cell-niche interaction. Biochem Biophys Res Commun 363:578–583
16. Katayama Y, Battista M, Kao WM, Hidalgo A, Peired AJ, Thomas SA, Frenette PS (2006) Signals from the sympathetic nervous system regulate hematopoietic stem cell egress from bone marrow. Cell 124:407–421
17. Kawasaki ES (2004) Microarrays and the gene expression profile of a single cell. Ann N Y Acad Sci 1020:92–100
18. Kiel MJ, Morrison SJ (2008) Uncertainty in the niches that maintain haematopoietic stem cells. Nat Rev Immunol 8:290–301
19. Kollet O, Dar A, Shivtiel S, Kalinkovich A, Lapid K, Sztainberg Y, Tesio M, Samstein RM, Goichberg P, Spiegel A et al (2006) Osteoclasts degrade endosteal components and promote mobilization of hematopoietic progenitor cells. Nat Med 12:657–664
20. Kurimoto K, Saitou M (2010) Single-cell cDNA microarray profiling of complex biological processes of differentiation. Curr Opin Genet Dev 20:470–477
21. Kurimoto K, Yabuta Y, Ohinata Y, Saitou M (2007) Global single-cell cDNA amplification to provide a template for representative high-density oligonucleotide microarray analysis. Nat Protoc 2:739–752
22. Lapidot T, Dar A, Kollet O (2005) How do stem cells find their way home? Blood 106:1901–1910
23. Lu R, Markowetz F, Unwin RD, Leek JT, Airoldi EM, MacArthur BD, Lachmann A, Rozov R, Ma'ayan A, Boyer LA et al (2009) Systems-level dynamic analyses of fate change in murine embryonic stem cells. Nature 462:358–362
24. Mendez-Ferrer S, Lucas D, Battista M, Frenette PS (2008) Haematopoietic stem cell release is regulated by circadian oscillations. Nature 452:442–447
25. Mendez-Ferrer S, Michurina TV, Ferraro F, Mazloom AR, Macarthur BD, Lira SA, Scadden DT, Ma'ayan A, Enikolopov GN, Frenette PS (2010) Mesenchymal and haematopoietic stem cells form a unique bone marrow niche. Nature 466:829–834
26. Moore KA, Lemischka IR (2006) Stem cells and their niches. Science 311:1880–1885
27. Morrison SJ, Spradling AC (2008) Stem cells and niches: mechanisms that promote stem cell maintenance throughout life. Cell 132:598–611
28. Nakamura Y, Arai F, Iwasaki H, Hosokawa K, Kobayashi I, Gomei Y, Matsumoto Y, Yoshihara H, Suda T (2010) Isolation and characterization of endosteal niche cell populations that regulate hematopoietic stem cells. Blood 116:1422–1432
29. Naveiras O, Nardi V, Wenzel PL, Hauschka PV, Fahey F, Daley GQ (2009) Bone-marrow adipocytes as negative regulators of the haematopoietic microenvironment. Nature 460:259–263
30. Nilsson SK, Johnston HM, Whitty GA, Williams B, Webb RJ, Denhardt DT, Bertoncello I, Bendall LJ, Simmons PJ, Haylock DN (2005) Osteopontin, a key component of the hematopoietic stem cell niche and regulator of primitive hematopoietic progenitor cells. Blood 106:1232–1239
31. Qian H, Buza-Vidas N, Hyland CD, Jensen CT, Antonchuk J, Månsson R, Thoren LA, Ekblom M, Alexander WS, Jacobsen SEW (2007) Critical role of thrombopoietin in maintaining adult quiescent hematopoietic stem cells. Cell Stem Cell 1:671–684

32. Rieger MA, Hoppe PS, Smejkal BM, Eitelhuber AC, Schroeder T (2009) Hematopoietic cytokines can instruct lineage choice. Science 325:217–218
33. Sacchetti B, Funari A, Michienzi S, Di Cesare S, Piersanti S, Saggio I, Tagliafico E, Ferrari S, Robey PG, Riminucci M et al (2007) Self-renewing osteoprogenitors in bone marrow sinusoids can organize a hematopoietic microenvironment. Cell 131:324–336
34. Singh D, Febbo PG, Ross K, Jackson DG, Manola J, Ladd C, Tamayo P, Renshaw AA, D'Amico AV, Richie JP et al (2002) Gene expression correlates of clinical prostate cancer behavior. Cancer Cell 1:203–209
35. Spurgeon SL, Jones RC, Ramakrishnan R (2008) High throughput gene expression measurement with real time PCR in a microfluidic dynamic array. PLoS One 3:e1662
36. Stahlberg A, Bengtsson M (2010) Single-cell gene expression profiling using reverse transcription quantitative real-time PCR. Methods 50:282–288
37. Stier S, Ko Y, Forkert R, Lutz C, Neuhaus T, Grunewald E, Cheng T, Dombkowski D, Calvi LM, Rittling SR et al (2005) Osteopontin is a hematopoietic stem cell niche component that negatively regulates stem cell pool size. J Exp Med 201:1781–1791
38. Sugiyama T, Kohara H, Noda M, Nagasawa T (2006) Maintenance of the hematopoietic stem cell pool by CXCL12-CXCR4 chemokine signaling in bone marrow stromal cell niches. Immunity 25:977–988
39. Thoren LA, Liuba K, Bryder D, Nygren JM, Jensen CT, Qian H, Antonchuk J, Jacobsen SE (2008) Kit regulates maintenance of quiescent hematopoietic stem cells. J Immunol 180:2045–2053
40. Thorsen T, Maerkl SJ, Quake SR (2002) Microfluidic large-scale integration. Science 298:580–584
41. Van Gelder RN, von Zastrow ME, Yool A, Dement WC, Barchas JD, Eberwine JH (1990) Amplified RNA synthesized from limited quantities of heterogeneous cDNA. Proc Natl Acad Sci U S A 87:1663–1667
42. Warren L, Bryder D, Weissman IL, Quake SR (2006) Transcription factor profiling in individual hematopoietic progenitors by digital RT-PCR. Proc Natl Acad Sci U S A 103:17807–17812
43. Wilson A, Laurenti E, Oser G, van der Wath RC, Blanco-Bose W, Jaworski M, Offner S, Dunant CF, Eshkind L, Bockamp E et al (2008) Hematopoietic stem cells reversibly switch from dormancy to self-renewal during homeostasis and repair. Cell 135:1118–1129
44. Wilson A, Trumpp A (2006) Bone-marrow haematopoietic-stem-cell niches. Nat Rev Immunol 6:93–106
45. Yin T, Li L (2006) The stem cell niches in bone. J Clin Invest 116:1195–1201
46. Yoshihara H, Arai F, Hosokawa K, Hagiwara T, Takubo K, Nakamura Y, Gomei Y, Iwasaki H, Matsuoka S, Miyamoto K et al (2007) Thrombopoietin/MPL signaling regulates hematopoietic stem cell quiescence and interaction with the osteoblastic niche. Cell Stem Cell 1:685–697

Chapter 2
Cell Fate Regulatory Networks

Sonya J. Ridden and Ben D. MacArthur

Abstract Cell fate decisions are controlled by intrinsically complex molecular regulatory networks, involving a wide variety of protein–protein and protein-DNA interactions. Due to this complexity, it is difficult to understand molecular regulation of cell fate at the 'systems' level. In this chapter we discuss mathematical modeling of cell fate regulatory networks, and explain some ways in which mathematical techniques may be used to elucidate the essential molecular mechanisms that underly cell fate determination. We consider both Boolean networks and ordinary differential equation (ODE) models. We give an illustrative worked example of an ODE model of a simple irreversible molecular switch due to an self-enhancing positive feedback loop and discuss how various commonly-occurring network 'motifs' can give rise to certain well-defined dynamics, including switches and oscillators. We conclude with some words on the role of stochasticity in cell fate regulatory networks.

Keywords Regulatory network • Cell fate • Mathematical model • Network motif

2.1 Introduction

A proper understanding of the molecular basis of cell fate decisions (including self-renewal, differentiation, apoptosis and quiescence for instance) is of critical importance in understanding both healthy and pathological tissue turnover. However,

S.J. Ridden
Institute for Complex Systems Simulation, School of Mathematics, University of Southampton, University Road, Southampton SO17 1BJ, UK
e-mail: S.J.Ridden@soton.ac.uk

Ben D. MacArthur (✉)
Faculty of Medicine and School of Mathematics, Institute for Life Sciences, University of Southampton, Southampton, United Kingdom
e-mail: B.D.MacArthur@soton.ac.uk

A. Ma'ayan and Ben D. MacArthur (eds.), *New Frontiers of Network Analysis in Systems Biology*, DOI 10.1007/978-94-007-4330-4_2,

molecular regulation of cell fate is complex, and depends upon numerous context specific details. Nevertheless, the search for general principles underpinning cell fate is the subject of significant current research interest, and has been for at least the last 60 years.

An early significant attempt to understand cell fate was presented in the 1950s by the developmental biologist Conrad Waddington when he introduced the notion of the 'epigenetic landscape' as a qualitative picture of development [48]. Waddington imagined the specification of different cell types occurring as a ball rolling down sloping channels in a landscape consisting of hills and valleys. As the ball (representing the cell) rolls down the hill, it reaches a point at which the channel splits in two, forcing the cell to chose between the different valleys. The downhill motion of the ball represents how the process moves inexorably forward in developmental time, while differentiation is controlled by the hills, which act as barriers separating the landscape into distinct valleys. This intuitive metaphor for development and the discrete nature of cell fates was particularly insightful given that relatively little was known about protein synthesis prior to the discoveries of Watson and Crick, whose work on the structure of DNA was also published throughout the 1950s [7, 50].

The increasing use of high-throughput molecular biology technology has generated a wealth of data, which, along with the development of computational network inference algorithms, is making it increasingly clear that cell fate decisions are not regulated by individual regulatory pathways, but rather by complex networks of intersecting pathways consisting of protein–protein and protein-DNA interactions that incorporate both transcriptional regulation and signaling mechanisms. However, this modern notion of molecular regulatory networks was envisioned at least as early as the 1960s. For instance, the molecular biologists Jaques Monod and Francois Jacob were early proponents of the notion that cell fates correspond to dynamically stable states of underlying molecular regulatory networks [32]. They note that microbial regulatory elements '... could be connected into a wide variety of 'circuits' endowed with any desired degree of stability'. Similarly (and also in the 1960s) the theoretical biologist Stuart Kauffman envisioned cell fates as arising from the dynamics of *complex* genetic regulatory networks [19].

These two, apparently disparate, pictures of cell fate are, in fact, closely related and much recent work has sought to unite Waddington's vision of the epigenetic landscape with the dynamics of genetic regulatory networks [28, 49] (we will discuss connections between these views in Sect. 2.3 of this chapter). However, after 60 years of study, the molecular complexity of cell fate regulation is only just becoming clear and the dynamics of the cell fate regulatory networks that underlie development are still the subject of considerable research interest.

2.2 Regulatory Networks

In mathematical terms a regulatory network is a set of a set of interconnected components, called *nodes* – which represent the molecular entities involved (generally genes and proteins) – along with connections between them called *links*.

Links represent interactions between molecular components and can be *directed* or *undirected.* For instance, much attention has been paid to using high-throughput experimental techniques to identify physical protein–protein interactions (see elsewhere in the present volume for more details). These data allow the inference of protein–protein interaction (PPI) networks, using a combination of experimental methods [23, 40, 47] and reverse engineering using computational techniques [2, 13, 34, 44]. In PPIs the nodes represent proteins and links represent physical interactions between proteins (i.e. binding). In this case, the links have no specific orientation: if protein A interacts with protein B then B also interacts with A. PPI networks are therefore undirected (see Fig. 2.3a, for a simple example). Although PPI networks map out the physical interactions between proteins – and thus the *possible* protein complexes which may form – they do not incorporate the consequences of these physical interactions, such as the induction or repression of gene expression by multi-protein complexes. These interaction effects can be represented in the form of a transcriptional regulatory network and much attention has also been paid to determining the structure and function of transcriptional regulatory networks [37]. Transcription factors (TFs) are nuclear proteins that, by binding to genomic regulatory regions and interacting with RNA polymerase II, activate or repress gene expression. In transcriptional regulatory networks the nodes are TFs and the links represent regulation of gene expression by upstream TFs. Unlike physical interactions between proteins there is a definite orientation to transcriptional regulatory interactions: if TF A regulates the transcription of TF B, it is not necessarily (or even generally) true that B regulates transcription of A. Consequently, transcriptional regulatory networks are directed.

In addition to protein–protein interactions and the action of transcription factors, gene expression is also regulated by epigenetic regulatory mechanisms, such as histone acetylation and DNA methylation, and signalling networks [21]. Thus, cell fate decisions are guided by the integrated action of protein–protein interaction networks, transcriptional regulatory networks, epigenetic regulators and signalling networks. The sheer number of molecular components in these integrated networks makes understanding them a formidable task. One approach that has proven useful in deciphering this complexity is the use of mathematical and computational models.

2.3 Modeling Regulatory Networks

The essential aim of mathematical models of regulatory networks is to convert known structural information (in the form of experimentally-derived protein–protein and protein-DNA interactions, for instance) into a set of equations that describe how molecular expression levels change over time as a result of these interactions, which may be solved in order to reproduce observed dynamics and make novel predictions concerning cell behavior. Thus, cell fate regulatory networks may be viewed as *dynamical systems* and simulations based on mathematical models can be used to explore how molecular regulatory interactions give rise to characteristic patterns of cell behavior. Mathematical models have been successfully combined with experimental data to study a wide variety of cell fate decisions [11, 14].

A cell's state may be described by its gene (or protein) expression profile, for instance as assessed by single cell expression profiling (see elsewhere in this volume and [25]). If there are N genes in the regulatory network, the expression profile is a set of N values, each of which indicates the level of expression of one gene. Since expression profiles are dynamic, we denote these values $x_i(t)$, where t represents time and $i = 1, \ldots, N$ indexes the genes. This set of numbers represents the state of the cell and is typically written in the form of a vector

$$\mathbf{x}(t) = [x_1(t), x_2(t), \ldots, x_N(t)],$$

called the *state vector*. The set of all possible genetic configurations is called the *state space*. The regulatory interactions between the components govern the *trajectory* of the state vector over time in the state space. From experiment it is clear that not all points in the state space (that is, all genetic profiles) are equal: some patterns of gene expression are only transiently expressed, while others will be expressed in a stable and sustained manner. In the language of dynamical systems, such stable states – which are robust to small perturbations – are known as *attractors*. In fact, attractors need not simply be stable static genetic configurations in which the rate of production is balanced by the rate of degradation: they may also be dynamic, corresponding to self-sustaining dynamic patterns of expression including stable oscillations.

2.3.1 Boolean Networks

In Boolean networks, gene expression is described by a binary variable that is either 'on' (active, 1) or 'off' (inactive, 0), so that each entry in the state vector takes the value 0 or 1 and the state space consists of 2^N states. For synchronous Boolean networks, time proceeds in discrete steps and the state of each gene is updated simultaneously, at every step. The state of each gene at time $t + 1$ is related to the states of its upstream regulators at the previous time step using Boolean functions (such as AND, OR, NOT, NAND and NOR). For example, if a gene with two inputs is governed by the Boolean AND function, then the state of the first input and of the second input must both be 1 ('on') at time t for the value of the output gene to equal 1 ('on') at time $t + 1$. If either of the upstream regulators are 'off' (0) at time t, then the target gene will also be 'off' at time $t + 1$. Alternatively, if a gene with two inputs is governed by the Boolean OR function, then the state of the first input or the second input must be 1 at time t for the value of the output gene to equal 1 at time $t + 1$. In this case, if both the upstream regulators are 'off' at time t then the target gene will also be 'off' at time $t + 1$.

Boolean networks were proposed as early models of cell fate regulatory networks by Stuart Kauffman in the 1960s [19] and have since attracted a lot of attention as models of complex molecular regulatory systems [20, 26, 39, 42]. Kauffman's aim was to explore the conditions required for large, biologically plausible, complex networks to give rise to dynamics similar to those exhibited by real genetic regulatory networks. By generating a large number of random Boolean networks (in which each node is connected to a fixed number of randomly selected nodes with randomly assigned Boolean functions) Kauffman investigated the effect of local network properties, such as the number of inputs received by each gene and the type of Boolean functions used to regulate expression, on global dynamics. A central finding of this work was the surprising conclusion that, given certain conditions, large regulatory networks can give rise to relatively few attractor states [19, 20]. Since the number of cell 'types' in the body is considerably less than the number of possible genetic configurations, this led Kauffman to propose that cell types correspond to attractors of high-dimensional regulatory networks [19, 20]. Experimental evidence supporting Kauffman's attractor hypothesis was recently provided by Huang et al. [16]. They showed that two biochemically distinct stimuli (the solvent DMSO and the hormone ATRA) were both able to trigger neutrophil differentiation in human promyelocytic cells. However, using a microarray time-series experiment they demonstrated that these stimuli did not guide differentiation in the same way: initially the two stimuli triggered divergence patterns of gene expression. Nevertheless, over time, the two different trajectories ultimately converged the same differentiated neutrophil state, indicating the presence of an attractor.

The most obvious criticism of Boolean networks, is that, in reality, gene expression levels are continuous, yet Boolean models quantize expression into discrete states. Furthermore, changes in gene expression do not occur discretely or synchronously and there can be significant differences in Boolean network dynamics if nodes are updated randomly or asynchronously [12]. A number of alternative formulations of Boolean networks have been considered to address these issues [9, 12, 38]. However, a more natural approach when considering continuous changes in gene expression to use ordinary differential equation models.

2.3.2 Rate Equations

Ordinary differential equations (ODEs) are the most common formalism used to model dynamical systems in science and engineering, and are widely used to model genetic regulatory systems [45, 46]. In the context of molecular regulatory networks ODEs are used to describe how the rates of production and decay of a gene product are determined by the concentrations of the upstream products that regulate its expression. Thus, ODE models consist of *rate equations* of the form

$$\frac{dx_i}{dt} = f_i(x_1, x_2, \ldots, x_N), \tag{2.1}$$

where x_i denotes the concentration of product i and the subscript i indexes the components of the network ($i = 1, \ldots N$). Equation 2.1 consists of N coupled equations, with one equation for each molecular component in the network (which may include both intracellular and extracellular molecules).

As a simple illustrative example consider the following model, a variation of that given in [3], of an irreversible molecular switch. Let x denote the concentration of a transcription factor X which activates its own transcription, and decays with constant half-life. Assuming that transcription is the rate-limiting step in protein production, the following ODE models this situation,

$$\frac{dx}{dt} = \frac{\alpha x^m}{K^m + x^m} - \beta x. \tag{2.2}$$

The first term on the right-hand side of this equation is a *Hill* function which determines the rate of production of X. It's value ranges from 0 to α and increases in a continuous manner as the concentration of X increases. Thus, α is the maximum rate of production of X (α has units of concentration per unit time). The parameter K is the concentration at which the half-maximum rate of production occurs, and the Hill coefficient m determines the gradient of the Hill function. The second term on the right-had-side of Eq. 2.2 says that X decays linearly (with constant half-life). The parameter β is the decay rate constant and has units t^{-1}.

Equation 2.2 has four free parameters (α, β, K and m). However, before beginning analysis it is convenient to rescale the system in order to aggregate these parameters and simplify calculations. We do this by introducing the scalings $x = \frac{\alpha}{\beta} y$ and $t = \frac{1}{\beta}\tau$, where y and τ are *dimensionless* variables. Equation 2.2 may then be expressed in dimensionless form:

$$\frac{dy}{d\tau} = \frac{y^m}{\gamma^m + y^m} - y. \tag{2.3}$$

Equation 2.3 behaves qualitatively in exactly the same way as Eq. 2.2. However, by writing the system in dimensionless form it becomes clear that the dynamics are governed not by four free parameters, but rather by m and a single 'lumped' dimensionless parameter $\gamma = \frac{K\beta}{\alpha} \geq 0$ which may be thought of, phenomenologically, as governing the strength of the auto-activatory feedback loop: at $\gamma = 0$ the feedback loop is fully active and production of X occurs at its maximal rate, while as $\gamma \to \infty$ feedback is inhibited and production of X tends to zero. The size of γ, and thus the strength of the feedback loop, may vary depending on internal or external signals and the dynamics of the system change as γ is varied.

In the absence of time-delays and noise, sustained oscillations (or more exotic stable dynamic behavior) can only occur when there is more than one species present. Since this model is one-dimensional we do not need to look for these here, so we look for *fixed-point* equilibrium solutions. At fixed-points the rate of change of the concentration of X is zero (i.e. production = degradation and $dy/d\tau = 0$) and the concentration of X remains constant. Therefore, we can calculate the values of the fixed-points by setting the right hand side of Eq. 2.3 equal to zero and solving for y. Fixed points, denoted y_*, are therefore solutions to a polynomial of degree $m + 1$,

$$y_*^{m+1} - y_*^m + \gamma^m y_* = 0. \tag{2.4}$$

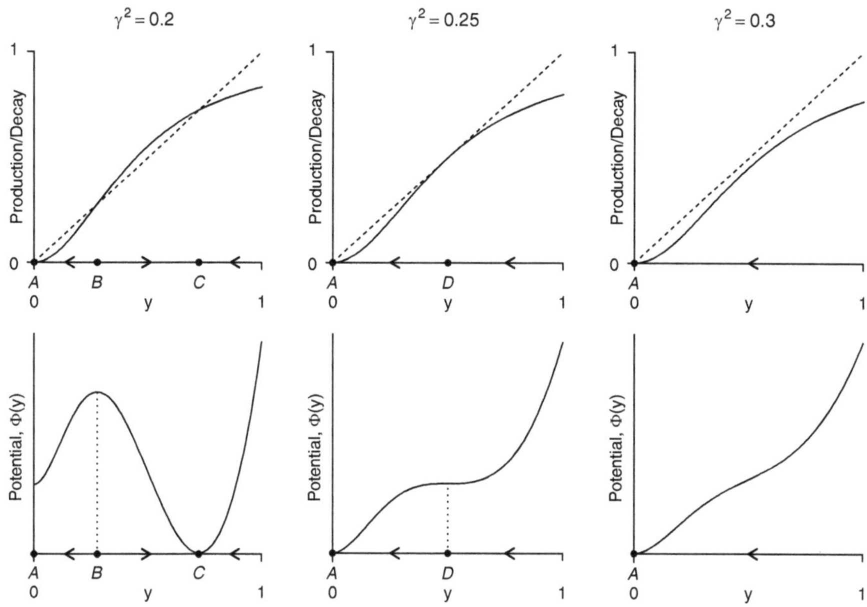

Fig. 2.1 (*Top row*) The Hill production function $\frac{y^2}{\gamma^2+y^2}$ (*solid line*), and degradation function y (*dashed line*), for $\gamma^2 = 0.2, 0.25, 0.3$. The filled circles mark the fixed-points, labelled A, B, C and D. The arrows on the *horizontal axes* indicate the direction of flow between the fixed-points. (*Bottom row*) The corresponding potential functions. The system exhibits bistability for $0 < \gamma^2 < 0.25$

For simplicity let's explore the case $m = 2$. Fixed-points are solutions of a cubic equation, and, in general, there are three distinct solutions. Figure 2.1 shows how the production and decay terms intersect (and thus the positions of the fixed-points change) as γ is varied. The origin ($y_* = 0$), labelled A, is always a fixed-point, regardless of the value of γ.[1] For small γ there are two more fixed-points,

$$y_* = \frac{1 \pm \sqrt{1 - 4\gamma^2}}{2}, \tag{2.5}$$

denoted B and C Fig. 2.1. As γ increases, B and C approach each other and eventually coalesce at the point D ($= 0.5$), when $\gamma^2 = 0.25$ and the production and decay rates intersect tangentially. As γ increases further ($\gamma^2 > 0.25$), the only remaining real fixed point is A. Thus, as the value of γ changes, fixed-points appear and disappear. Additionally, their stability changes. The parameter values at which these qualitative changes in dynamics occur are called *bifurcation points*.

The plots in Fig. 2.1 allow us to determine the stability of the identified fixed-points. When the value of the production function is greater than that of the

[1] In fact, from Eq. 2.4 this is always the case for m a positive integer.

degradation function, the concentration of X increases and the flow along the horizontal axis is to the right. Conversely, when the value of the production function is less than that of the degradation function, the concentration of X decreases and the flow is to the left. The arrows marked on the horizontal axes indicate the direction of flow. So, for $\gamma^2 < 0.25$ the direction of flow is towards A and C and they are *stable*, while the flow is away from B and it is *unstable*. In this case, the system is *bistable*, since it supports two coexisting stable fixed-points. The unstable fixed-point B separates the two stable states, like a hill between two valleys. Consequently, initial conditions in the region $0 < x < B$, are drawn toward A over time, while initial values in the region $x > B$ are drawn to C over time. We say that $0 < x < B$ is the *basin of attraction* of A, and $x > B$ is the basin of attraction of C. Similar analysis shows that at $\gamma^2 = 0.25, D$ is an unstable fixed-point and the only stable fixed-point is A. Hence, the system exhibits bistability when $0 < \gamma^2 < 0.25$.

This system is an irreversible molecular switch in the following sense. Suppose we are operating in the bistable regime, and we start in the vicinity of C. We now gradually increase γ until $\gamma^2 = 0.25$. At this point, C loses stability and the system will transition to A, the only stable fixed-point left. If we now decrease γ back below the critical point $\gamma^2 = 0.25$ the system will not switch back to C, but rather will remain fixed at A since A is stable for all γ. Thus, once the transition from C to A has been made it cannot be reversed: the switch is one-way. It is not always true that switches are irreversible. They may be perfectly reversible or exhibit a delayed return response known as *hysteresis* [22].

As another example, let's consider the case $m = 1$. Fixed-points are then $A = 0$ and $B = 1 - \gamma$.[2] Figure 2.2 shows the intersections of the production and degradation functions for $\gamma = 0.2$. Here A is unstable and B is stable and, in fact, this is the case for all $0 < \gamma < 1$. At $\gamma = 1$, A and B coalesce and exchange stability (the origin becomes stable), and this remains the case for $\gamma \geq 1$. Thus, the system does not exhibit bistability for any value of γ when $m = 1$. The reason for this is that the production function must be sigmoidal to intersect the degradation function at the three points needed for the system to exhibit bistability, and this will only happen when $m > 1$. This illustrates a key point: some kind of nonlinearity is usually needed – in the form of cooperative binding, for instance – to support bistability, although this is not always the case [27].

The simple model outlined above allows us to see how Waddington's idea of the epigenetic landscape connects to the notion of attractors of dynamical systems. We first observe that the right-hand-side of Eq. 2.3 may be written as the negative gradient of a *potential function* $\Phi(y)$

[2] Note that for $\gamma > 1, B < 0$, which physically unrealistic. However, starting from a non-negative initial condition the dynamics are confined to $y \geq 0$ since the origin is a nullcline of the system. So, for biologically realistic initial conditions, negative values for the concentration are not obtained.

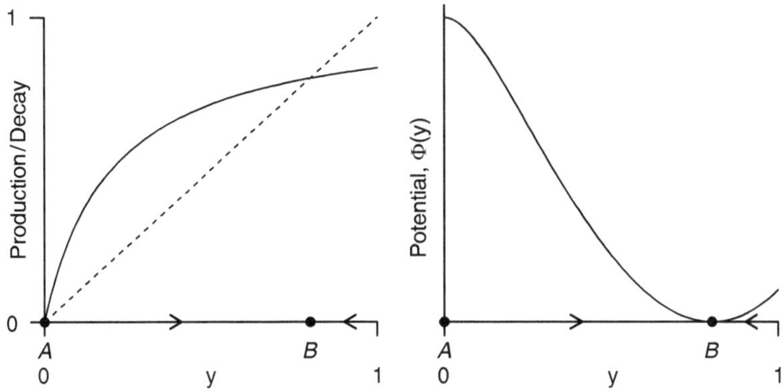

Fig. 2.2 (*Left*) Plot of the production function $\frac{y}{\gamma+y}$ (*solid line*), and degradation function y (*dashed line*), for $\gamma = 0.2$. The filled circles mark the fixed-points, labelled A and B. The arrows on the *horizontal axes* indicate the direction of flow between the fixed-points. (*Right*) The corresponding potential function. The system does not exhibit bistability for any value of γ when $m = 1$ because the production function is not sigmoidal

$$\frac{dy}{d\tau} = \frac{y^m}{\gamma^m + y^m} - y = -\frac{d\Phi}{dy}.$$

In the case that $m = 2$ we obtain

$$\Phi(y) = \gamma \arctan\left(\frac{y}{\gamma}\right) + \frac{y^2}{2} - y,$$

while for $m = 1$ we have

$$\Phi(y) = \gamma \log(\gamma + y) + \frac{y^2}{2} - y.$$

The key reason for making this substitution is that the stable fixed-points of the system correspond to minima ('valleys') of $\Phi(y)$, while unstable fixed-points correspond to maxima ('hills') of $\Phi(y)$, thus providing a 'landscape' for the dynamics. More precisely, the system always tries to minimize $\Phi(y)$ thus making quantitative Waddington's qualitative notion of a ball rolling down a ruffled landscape. In one dimension this technique may always be performed. However, in higher dimensions it is not always possible to write the dynamics in terms of the gradient of a scalar potential function. Nevertheless, this notion is useful for examining gradient systems, and makes explicit the connection between dynamics of regulatory networks and Waddington's epigenetic landscape. Figures 2.1 and 2.2 show the structure of the potential function of our model for different values of γ and m.

Rate equation models and Boolean networks capture the dynamic behavior of regulatory networks at different levels. Boolean networks allow us to model large-scale regulatory networks, in order to elucidate 'systems-level' organizational principles. However Boolean networks assume that each gene can only take a finite

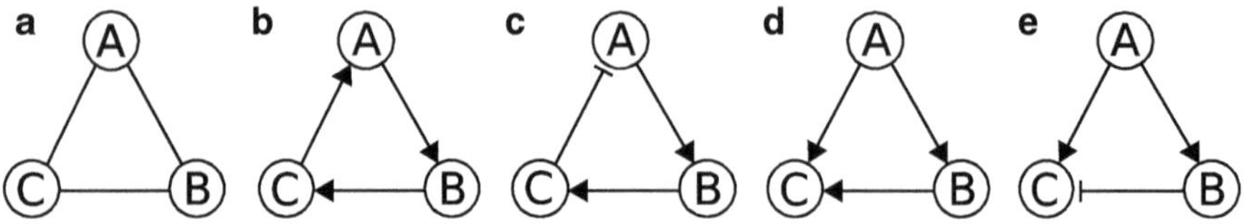

Fig. 2.3 Examples of network motifs consisting of three genes: (**a**) an undirected cycle; (**b**) a self-enhancing positive feedback loop; (**c**) a negative feedback loop; (**d**) a coherent feedforward loop; (**e**) an incoherent feedforward loop. Activation of gene expression is represented by an *arrow*, and inhibition by a T-bar

number of discrete states, and so are not able to model regulatory networks in fine detail. Rate equation models, on the other hand, allow us to study regulatory systems in fine detail, and so are usually limited (both computationally, and in the amount of biological data needed to build them) to relatively small systems.

2.3.3 Network Motifs

Although complex in their overall structure [4, 8], molecular regulatory networks often contain certain types of small subnetworks at frequencies higher than expected by chance [36, 37], suggesting that these structural building blocks – or *motifs* as they are known – may perform specific regulatory functions [37]. In the previous section we considered a simple model of a switch due to a nonlinear positive feedback loop. Here, we shall consider further the dynamics of three types of commonly occurring, and dynamically significant, motifs: positive feedback, negative feedback and feed-forward loops. Examples of these motifs are shown in Fig. 2.3.

2.3.3.1 Positive and Negative Feedback Loops: Switches, Homeostasis and Oscillators

The essential underlying property of Eq. 2.3 which formed the basis of the switch we observed was bistability. In general, a necessary condition for bistability (or multistability[3]) is the presence of a *positive feedback loop* somewhere in the underlying regulatory network [43, 51, 52]. A feedback loop is a closed path in a network from a node back to itself in which each intermediary node is visited once. A positive feedback loop is one in which the net effect of the entire loop is positive: using the convention that activating links are denoted by $+\,1$ and inhibiting links by $-\,1$, a positive feedback loop is one in which the product of the link-signs is $+\,1$. A positive feedback loop of length three is shown in Fig. 2.3b: in this case, all links are positive and the sign of the loop is therefore $(+1) \times (+1) \times (+1) = +1$.

[3] A dynamical system that supports more than one coexisting attractor is said to be *multi-stable*.

Since all the links are positive, this example feedback loop is *self-enhancing*. Once activated (perhaps in response to a transient activating signal), a self-enhancing feedback loop maintains the expression of all the genes in the loop. As in our worked example, this kind of positive feedback often provides the molecular basis for irreversible switches which, by initiating all-or-none cell fate decisions, are important in cellular differentiation and development [24, 33, 46].

If some of the links in the feedback loop are negative (and there must be an even number of negative links in a positive feedback loop) then alternative dynamics may be exhibited. For instance, a common motif observed in genetic regulatory networks is a pair of mutually repressing genes (this is a positive feedback loop of length two since its sign is $(-1) \times (-1) = +1$).This motif gives rise to a *toggle switch* [11], since it allows the cell to switch (or toggle) between two different fates in response to a transient signal. The key property of this switch is that the mutual repression between the two genes does not allow them both to be co-expressed at high levels. Consider two attractors: A (in which the first gene is active and, due to the repression exerted by gene 1, the second is inactive), B(in which the second gene is active and, due to the repression exerted by gene 2, the first is inactive), and a signal of magnitude S. Now suppose that if the system starts in the vicinity of A it can be driven out of the basin of attraction of A and into the basinof attraction of B in response to a signal that exceeds a critical value S_{HIGH} in magnitude. Similarly, suppose that if the system starts in the vicinity of$B$ it can switch to A if the signal magnitude falls below S_{LOW}. For intermediate signal magnitudes ($S_{LOW} < S < S_{HIGH}$), the system is bistable since it admits two coexisting attractors. Thus, varying S allows the system to toggle between the two alternative attractor states. A particularly elegant example was given by Gardner et al. [11]. They constructed a synthetic toggle switch in *Escherichia coli*, based on the predictions of a simple mathematical model, and showed that cell fate could becontrolled using this switch by introducing a transient chemical or thermal signal.

A negative feedback loop is a feedback loop in which the net effect of the entire loop is negative (one in which the product of the link-signs is -1). An example of a negative feedback loop is shown in Fig. 2.3c: in this case, the sign of the loop as a whole is $(+1) \times (+1) \times (-1) = -1$. Homeostasis (the maintenance of a constant internal state despite environmental variations) may be maintained by negative feedback, since negative feedback suppresses fluctuations [35]. Similarly, negative feedback loops can, by introducing time-delays [15] and associated over- and under-compensation in gene expression, give rise to self-sustaining oscillatory behavior [46]. An example of a synthetic three-gene negative feedback loop that generates self-sustained oscillations in protein levels in *Escherichia coli* was presented by Elowitz and Leibler [10]. They showed that this system, named the *repressilator*, can function as a biological clock by inducing periodic bursts of protein synthesis.

2.3.3.2 Feedforward Loops: Persistence Detectors and Pulse Generators

Feedforward loops occur when a source gene regulates the expression of a target gene through two different paths. Figure 2.3d, e show examples of three-node feedforward loops. In both these cases, *A* regulates *C* both directly and indirectly via *B*. Feedforward loops are common in molecular regulatory networks, including the transcriptional regulatory networks of *Escherichia coli* and yeast [29, 30, 37] and other organisms [5, 18, 31]. Each of the regulatory interactions in the feedforward loop can be either be activating or repressing: if both paths in the feedforward loop have the same overall sign (both activating or both inhibiting) then the feedforward loop is said to be *coherent*, otherwise it is *incoherent*. The feedforward loop in Fig. 2.3d is coherent because the sign of the direct path from *A* to *C* has the same sign (positive in this case) as the indirect path from *A* to *C* via *B*. The feedforward loop in Fig. 2.3e is incoherent because the sign of the direct path from *A* to *C* is positive, while the sign of the indirect path from *A* to *C* via *B* is negative.

Coherent and incoherent feedforward loops exhibit different dynamics. Coherent feedforward loops can filter out transient environmental fluctuations and act as persistence detectors [1, 29]. For instance, consider the three node feedforward loop in Fig. 2.3d. If expression of both *A and B* is needed to activate *C* (if *C* is regulated by an *AB*-dimer, for example) then an activating signal starting at *A* must persist long enough for the concentration of *B* to reach the activation threshold before *C* is activated. In contrast, when the activating signal is removed, and *A* is down-regulated, the expression of *C* also down-regulates without delay. If expression of *C* only requires expression of *A or B* (*A* and *B* regulate *C* independently), then the opposite effect is observed: there is no delay in activation of *C* after activation of *A*, but there is a delay in down-regulation of *C* when stimulation of *A* stops [1].

Incoherent feedforward loops can act as pulse generators. For example, consider the three node feedforward loop in Fig. 2.3e. Node *A* both directly activates *C* and indirectly represses *C* by activating the repressor *B*. Consequently, when a signal activates *A*, the production of *C* is also rapidly activated. However, over time, levels of *B* also accumulate until they reach the repression threshold. At this point production of *C* decreases and its concentration drops, resulting in pulse-like expression of *C*.

2.3.3.3 Does Structure Determine Function?

Although certain network structures can be associated with defined dynamics, caution should be exercised when determining the relationship between structure and function in more complex regulatory networks. A study performed by Ingram et al. [17] showed that the function of even very simple motifs cannot always be determined by their structure. The authors investigate the behavior of the bi-fan motif – in which the products of two source genes directly co-regulate the expression of two target genes – and find that there is no characteristic behavior for this motif: the bi-fan can exhibit a range of possible responses. Given that the bi-fan is only slightly more complex than a feedforward loop, the authors conclude that '... it is difficult to gain significant insights into biological function simply by considering the connection architecture of a gene network, or its decomposition into simple structural motifs'. They add that additional information, such as the values of the kinetic parameters, or experimental time series data are required to make inferences about network dynamics.

2.4 Conclusion

Since Waddington's early conceptualization of development much progress has been made towards elucidating the general principals of cell fate determination, and recent advances in high-throughput technologies are allowing us to dissect the complexity of cell fate regulatory networks in ever more detail. In this chapter we have briefly considered various approaches to mathematical modeling of the deterministic dynamics of cell fate regulatory networks. However, gene and protein expression are 'noisy' and molecular noise also plays a key role in determining cell fate by driving stochastic transitions between coexisting attractor states or excursions in excitable systems [6, 41] giving rise to 'nongenetic' heterogeneity in isogenic cellular populations. Stochastic models of networks are discussed further elsewhere inthis volume. Ultimately, a proper understanding of cell fate will necessitate elucidation of both stochastic and deterministic mechanisms. This will require integration of experimental and computational approaches and the collaboration of scientists from a wide range of disciplines.

Acknowledgements This work was supported by an EPSRC Doctoral Training Centre grant (EP/G03690X/1).

References

1. Alon U (2007) Network motifs: theory and experimental approaches. Nat Rev Genet 8(6):450–461
2. Bader J, Chaudhuri A, Rothberg J, Chant J (2004) Gaining confidence in high-throughput protein interaction networks. Nat Biotechnol 22(1):78–85
3. Becskei A, Séraphin B, Serrano L (2001) Positive feedback in eukaryotic gene networks: cell differentiation by graded to binary response conversion. EMBO J 20(10):2528–2535
4. Bolouri H, Davidson E (2002) Modeling transcriptional regulatory networks. BioEssays 24(12):1118–1129
5. Boyer L, Lee T, Cole M, Johnstone S, Levine S, Zucker J, Guenther M, Kumar R, Murray H, Jenner R, et al (2005) Core transcriptional regulatory circuitry in human embryonic stem cells. Cell 122(6):947–956
6. Chang HH, Hemberg M, Barahona M, Ingber DE, Huang S (2008) Transcriptome-wide noise controls lineage choice in mammalian progenitor cells. Nature 453:544–547
7. Crick F (1958) On protein synthesis. Symp Soc Exp Biol 12:138
8. Davidson E (2007) Gene regulation: gene control network in development. Annu Rev Biophys Biomol Struct 36:191–212
9. Dougherty E, Datta A, Sima C (2005) Research issues in genomic signal processing. IEEE Signal Process Mag 22(6):46–68
10. Elowitz M, Leibler S (2000) A synthetic oscillatory network of transcriptional regulators. Nature 403(6767):335–338
11. Gardner T, Cantor C, Collins J (2000) Construction of a genetic toggle switch in Escherichia coli. Nature 403(6767):339–342

12. Gershenson C (2004) Introduction to random boolean networks. In: Bedau M, Husbands P, Hutton T, Kumar S, Suzuki H (eds) Workshop and tutorial proceedings, ninth international conference on the simulation and synthesis of living systems (ALife IX), pp 160–173
13. Giot L, Bader J, Brouwer C, Chaudhuri A, Kuang B, Li Y, Hao Y, Ooi C, Godwin B, Vitols E, et al (2003) A protein interaction map of drosophila melanogaster. Science 302(5651):1727
14. Hasty J, McMillen D, Isaacs F, Collins J (2001) Computational studies of gene regulatory networks: in numero molecular biology. Nat Rev Genet 2(4):268–279
15. Heiden U (1979) Delays in physiological systems. J Math Biol 8(4):345–364
16. Huang S, Eichler G, Bar-Yam Y, Ingber D (2005) Cell fates as high-dimensional attractor states of a complex gene regulatory network. Phys Rev Lett 94(12):128, 701
17. Ingram P, Stumpf M, Stark J (2006) Network motifs: structure does not determine function. BMC Genomics 7(1):108
18. Iranfar N, Fuller D, Loomis W (2006) Transcriptional regulation of post-aggregation genes in dictyostelium by a feed-forward loop involving gbf and lagc. Dev Biol 290(2):460–469
19. Kauffman S (1969) Metabolic stability and epigenesis in randomly constructed genetic nets. J Theor Biol 22(3):437–467
20. Kauffman S (1993) The origins of order, vol 209. Oxford University Press, New York
21. Kholodenko B (2006) Cell-signalling dynamics in time and space. Nat Rev Mol Cell Biol 7(3):165–176
22. Kramer BP, Fussenegger M (2005) Hysteresis in a synthetic mammalian gene network. Proc Natl Acad Sci USA 102(27):9517–9522
23. Landgraf C, Panni S, Montecchi-Palazzi L, Castagnoli L, Schneider-Mergener J, Volkmer-Engert R, Cesareni G (2004) Protein interaction networks by proteome peptide scanning. PLoS Biol 2(1):e14
24. Laurent M, Kellershohn N (1999) Multistability: a major means of differentiation and evolution in biological systems. Trends Biochem Sci 24(11):418–422
25. Levsky JM, Shenoy SM, Pezo RC, Singer RH (2002) Single-cell gene expression profiling. Science 297(5582):836–840
26. Liang S, Fuhrman S, Somogyi R, et al (1998) Reveal, a general reverse engineering algorithm for inference of genetic network architectures. Pac Symp Biocomput 3:18–29
27. Lipshtat A, Loinger A, Balaban N, Biham O (2006) Genetic toggle switch without cooperative binding. Phys Rev Lett 96(18):188101
28. MacArthur B, Ma'ayan A, Lemischka I (2009) Systems biology of stem cell fate and cellular reprogramming. Nat Rev Mol Cell Biol 10(10):672–681
29. Mangan S, Alon U (2003) Structure and function of the feed-forward loop network motif. Proc Natl Acad Sci USA 100(21):11980
30. Milo R, Shen-Orr S, Itzkovitz S, Kashtan N, Chklovskii D, Alon U (2002) Network motifs: simple building blocks of complex networks. Science 298(5594):824
31. Milo R, Itzkovitz S, Kashtan N, Levitt R, Shen-Orr S, Ayzenshtat I, Sheffer M, Alon U (2004) Superfamilies of evolved and designed networks. Science 303(5663):1538
32. Monod J, Jacob F (1961) General conclusions: teleonomic mechanisms in cellular metabolism, growth, and differentiation. In: Cold spring harbor symposia on quantitative biology, vol 26. Cold Spring Harbor Laboratory Press, Cold Spring Harbor, p 389
33. Niswander L, Jeffrey S, Martin G, Tickle C (1994) A positive feedback loop coordinates growth and patterning in the vertebrate limb. Nature 371(6498):609–612
34. Rinner O, Mueller L, Hubálek M, Müller M, Gstaiger M, Aebersold R (2007) An integrated mass spectrometric and computational framework for the analysis of protein interaction networks. Nat Biotechnol 25(3):345–352
35. Savageau M (1974) Comparison of classical and autogenous systems of regulation in inducible operons. Nature 252(5484):546–549
36. Segal E, Shapira M, Regev A, Pe'er D, Botstein D, Koller D, Friedman N (2003) Module networks: identifying regulatory modules and their condition-specific regulators from gene expression data. Nat Genet 34(2):166–176

37. Shen-Orr S, Milo R, Mangan S, Alon U (2002) Network motifs in the transcriptional regulation network of Escherichia coli. Nat Genet 31(1):64–68
38. Shmulevich I, Dougherty E, Kim S, Zhang W (2002) Probabilistic Boolean networks: a rule-based uncertainty model for gene regulatory networks. Bioinformatics 18(2):261
39. Somogyi R, Sniegoski C (1996) Modeling the complexity of genetic networks: understanding multigenic and pleiotropic regulation. Complexity 1(6):45–63
40. Stelzl U, Worm U, Lalowski M, Haenig C, Brembeck F, Goehler H, Stroedicke M, Zenkner M, Schoenherr A, Koeppen S, et al (2005) A human protein–protein interaction network: a resource for annotating the proteome. Cell 122(6):957–968
41. Süel G, Garcia-Ojalvo J, Liberman L, Elowitz M (2006) An excitable gene regulatory circuit induces transient cellular differentiation. Nature 440:545–550
42. Szallasi Z, Liang S (1998) Modeling the normal and neoplastic cell cycle with realistic Boolean genetic networks: their application for understanding carcinogenesis and assessing therapeutic strategies. Pacific Symposium on Biocomputing, vol 3. Stanford University, Stanford, pp 66–76
43. Thomas R (1981) On the relation between the logical structure of systems and their ability to generate multiple steady states or sustained oscillations. Springer Series Syne 9:180–193
44. Tong A, Drees B, Nardelli G, Bader G, Brannetti B, Castagnoli L, Evangelista M, Ferracuti S, Nelson B, Paoluzi S, et al (2002) A combined experimental and computational strategy to define protein interaction networks for peptide recognition modules. Science 295(5553):321
45. Tyson J, Chen K, Novak B (2001) Network dynamics and cell physiology. Nat Rev Mol Cell Bio 2(12):908–916
46. Tyson J, Chen K, Novak B (2003) Sniffers, buzzers, toggles and blinkers: dynamics of regulatory and signaling pathways in the cell. Curr Opin Cell Biol 15(2):221–231
47. Uetz P, Giot L, Cagney G, Mansfield T, Judson R, Knight J, Lockshon D, Narayan V, Srinivasan M, Pochart P, et al (2000) A comprehensive analysis of protein–protein interactions in saccharomyces cerevisiae. Nature 403(6770):623–627
48. Waddington C (1957) The strategy of the genes: a discussion of some aspects of theoretical biology. Allen & Unwin, London
49. Wang J, Zhang K, Xu L, Wang E (2011) Quantifying the waddington landscape and biological paths for development and differentiation. Proc Natl Acad Sci USA 108:8257–8262
50. Watson J, Crick F (1953) Molecular structure of nucleic acids. Nature 171(4356):737–738
51. Xiong W, Ferrell J (2003) A positive-feedback-based bistable memory modulethat governs a cell fate decision. Nature 426(6965):460–465
52. Zhang P, Behre G, Pan J, Iwama A, Wara-Aswapati N, Radomska H, Auron P, Tenen D, Sun Z (1999) Negative cross-talk between hematopoietic regulators: GATA proteins repress PU. 1. Proc Natl Acad Sci USA 96(15):8705

Chapter 3
Noise in Regulatory Networks

Srinandan Dasmahapatra

Abstract Gene regulatory networks control the production of the principal functional units (proteins, RNA) in cells. The observation of variability in expression levels within isogenic populations of cells demands an understanding of how this variability changes our understanding of cellular behaviour inferred from bulk measurements. In particular, we focus on simple models of gene expression to identify the key processes and their relative durations that determine fluctuation propagation within regulatory networks. We also point out modes of noise reduction and phenotypic state switching that rely on the average dynamics of regulatory networks, as well as on spatial characteristics of their embedding within organisms.

Keywords Transcriptional regulation • Stochasticity • Phenotype switch • Expression variation • Dynamic models

3.1 Introduction

Variations lie at the core of biology, shaping evolution by contributing to the pool of choices available to natural selection. Mutations in the genomic nucleotide sequences give rise to one source of heritable variation resulting in novel proteins with downstream consequences. The variations that are considered in this chapter are those of quantity – the codes for messenger RNA and proteins remain unchanged, but the numbers of proteins or nucleic acid molecules of specific identity vary in isogenic settings. This shifts the focus of study from the presence or absence of particular genes or of interactions between molecular species to quantitative analyses that relate to the numbers of molecules and kinetics of their changes. The relation of

S. Dasmahapatra (✉)
Faculty of Physical and Applied Sciences, University of Southampton,
Southampton SO17 1BJ, UK
e-mail: sd@ecs.soton.ac.uk

A. Ma'ayan and Ben D. MacArthur (eds.), *New Frontiers of Network Analysis in Systems Biology*, DOI 10.1007/978-94-007-4330-4_3,

genotype through to protein levels which we take to be proxy for phenotype involves many stages each subject to regulation and variation, and which are parameterised by the rate or probability of these steps in regulatory networks wired together via interactions between proteins, genes and mRNA and proteolytic machinery. These mutual influences mediate genotype-phenotype mapping as they regulate the expression, transformation and degradation of proteins. The resulting evolution of these molecular numbers is viewed mathematically as stochastic dynamical systems. While many of these processes are local to the gene that is variably expressed, interactions with other gene products that are themselves fluctuating can add to the variations in expression levels. These properties will be explored in this chapter.

Once a heterogeneous isogenic population arises due to variations in the number of gene products, one might consider how this distribution is related to system-wide properties. On an evolutionary scale, this distribution is filtered by the degree to which an organism can survive and reproduce in a specific environment. While everyday intuitions may cast variability as detrimental to an organism, hence the term "noise," it is clear that what matters is the match between what is dynamically enabled by the system and the environment it inhabits. Since we factor out molecular identity from our considerations, this matching is studied in terms of spatial and temporal characteristics of the environment. For instance, the induction of competence – a state in which a non-proliferating bacterium, *B. subtilis* can absorb DNA fragments from its environment – has been shown to occur more frequently in the presence of noise in gene expression [33, 55]. A genetically modified strain that performs the same function but with a narrower range of expression has also been constructed to compare its viablity to that of the wild type. The relative fitness of the strains has been shown to depend on the temporal variability of its environment [12], with the more variable wild type strain performing better across varied environments.

The cell-to-cell variations in gene expression are measured by fluorescence based cell-sorting technologies. In many cases these distributions are bimodal (or even multimodal) suggesting the existence of multiple expression phenotypes in isogenic populations. This was first demonstrated in the pre-fluorescence technology era by Novick and Weiner [34] using inducer dilution experiments and more recently tracked in single cells experimentally using the Lac operon system [1, 38]. It has also been shown [3] that unicellular organisms can switch between these multiple phenotypic states which, depending on the time scale of environmental fluctuations, can enhance their chances for survival, a property linked to antibiotic resistance.

Phenotypic consequences of expression variability has also been shown in eukaryotes, particularly promoter-driven (involving TATA and TATA-less promoters) transcriptional variability in yeast which propagates downstream, affecting other genes, altering phenotype [9, 10]. The propagation of variable expression levels down developmental networks in multicellular organisms links expression noise in individual cells to cell-fate and tissue determining choices. For instance, enhanced variability of expression arises owing to a mutation in a gene encoding a transcription factor within a regulatory network of the digestive tract of *C. elegans* which reduces the degree of redundancy behind the expression level of a downstream gene. Consequently, only a fraction of the worms that possessed a mutation in the network displayed the corresponding phenotype [43] compared to the wild-type, demonstrating that a phenonmenon called incomplete penetrance in evolutionary biology can be traced to expression variations. Further manipulations within this network have pointed to the contribution of chromatin remodelling to this variability, adding further evidence in the multicellular setting to earlier studies in yeast [10, 44].

These observations highlight the significance of expression noise in regulatory networks for individual organisms as well as for evolutionary processes. Waddington's concept of canalization [62] was predicated on buffering of fluctuations in developmental processes whether they were generated via environmental effects or internally, suggesting a common pool for variation was biologically meaningful processes to draw upon [31]. Observed correlations between expression variations in response to experimental cues and expression divergences between related yeast species ([57, 58] and references therein) point to gene promoters as loci for both short-term responses relevant for cells as well as to evolutionary time scales, particularly with respect to TATA-box containing promoters and chromatin modifiers. It has also been suggested [18] that variation in gene expression in populations of cells is correlated with expression differences in isogenic cell populations. While these studies have been done on yeast cells, correlation of expression patterns between proteins have also been performed in other organisms; mammalian single-cell studies that have highlighted the presence of multiple dynamical expression patterns in clonal cell lines [14] as well as temporal correlations between expression fluctuations [51].

While the role of noise is considered significant in understanding the nature of phenotypes at a systemic level, in most of the better understood systems the explanans lies in constituent properties – individual gene expression, promoter characteristics, local chromatin modifications, the network dynamics of a few component genes. In this chapter we will review mathematically guided interpretations of the consequences of noise in gene expression which take into account various sources of variability. We will start in Sect. 3.2 with simple models of protein synthesis and degradation to illustrate variability at the level of a single gene [40], introducing Markov processes in continuous time as a modelling framework. In order to encompass further aspects of the complexity of events leading up to transcription initiation, such as mechanisms of nucleosome reassembly and chromatin remodelling [11], we move on to consider mutiple promoter states for transcriptional control [9, 10, 49] in Sect. 3.3. Tracing the variability in transcriptional control at the promoter to variability of transcription factor expression levels leads to considerations of propagation of noise in transcriptional networks. In Sect. 3.4 we set up the discussion of the sources and dispersal of variations, focussing on issues of local and spatial context and temporal scales, where the mathematics of random walks is introduced in order to guide intuition and modelling approaches. We also indicate how the propagation of noise within networks could leave signature traces of the dynamics of regulation that are accessible to experimental probes. In Sect. 3.5, we separate out the contributions of molecular concentrations as the average dynamics of a network from the fluctuations about that average. In a few examples we show how the dynamics of the averages offers a multiplicity of phenotypic states to individual cells and the role of noise appears to enable transitions within the organism between the available states.

3.2 Stochastic Processes and Variability

We first introduce a minimal mathematical model to show how the variability in protein numbers may be quantified, and then point out how further biological realism may be introduced into the modelling framework. The amount of specific protein in a cell is the result of balancing protein degradation against production events driven by the elaborate set of biochemical reactions that comprise gene expression. Both these processes are stochastic in nature since they rely on molecular interactions in an isothermal cellular environment. A convenient modelling framework for capturing these stochastic transitions is that of a Markov process in continuous time which accounts for changes in the probability $P(\boldsymbol{X}, t)$ of finding the system in state $\boldsymbol{X}$ at time t. The state $\boldsymbol{X}$ can be taken to include, say, the state of the promoter/enhancer of the protein-coding gene, the number of mRNA transcripts and the number of proteins expressed. It is of course, possible to add further layers of complexity corresponding to nuclear or cytoplasmic locator labels on proteins and transcripts, post-transcriptional and post-translational modification steps, numbers of RNA polymerases, ribosomes, availability of nucleoside triphosphates (NTP), and so on. In what we will consider in this chapter, mRNA and protein numbers and a discrete set of states determining transcription initiation will be the random variables to keep track of. These state variables get updated (from $\boldsymbol{X}'$ to $\boldsymbol{X}$ below) at each reaction step whose probability of occurrence depends on the current state of the system (the Markov memoryless property):

$$P(\boldsymbol{X}, t + \Delta t) = \sum_{\boldsymbol{X}'} P(\boldsymbol{X}, t + \Delta t | \boldsymbol{X}', t) P(\boldsymbol{X}', t), \tag{3.1}$$

where the conditional probability $P(\cdot\,|\,\cdot)$ captures all the information required to describe the system at time $t + \Delta t$ given the distribution at t. The continuous time assumption allows us to choose infinitesimal time intervals in which the probability of occurrence of two reaction steps vanishes at $O(\Delta t)$, rendering the reactions considered mutually exclusive and the corresponding probability updates additive over possible reactions. Let $a_j(\boldsymbol{X})\Delta t$ be the probability of occurrence of the jth of the M reaction channels which changes species numbers $\boldsymbol{X} \to \boldsymbol{X} + \boldsymbol{\nu}_j$, where the elements of $\boldsymbol{\nu}_j$ form the stoichiometry matrix ν_{ji}, a book-keeping device that keeps track of the changes in the population of species i due to reaction j. Then the mutually exclusive reactions update the probability of the state additively,

$$P(\boldsymbol{X}, t + \Delta t) = \sum_{j=1}^{M} P(\boldsymbol{X} - \boldsymbol{\nu}_j, t) a_j(\boldsymbol{X} - \boldsymbol{\nu}_j) \Delta t + \left(1 - \Delta t \sum_{j=1}^{M} a_j(\boldsymbol{X})\right) P(\boldsymbol{X}, t), \tag{3.2}$$

and the last term accounts for the probability of the state to remain unchanged.

The terms in (3.2) may be rearranged to yield the chemical master equation, which is often taken as a starting point for subsequent analysis:

$$\frac{\partial}{\partial t}P(\boldsymbol{X},t) = \sum_{j=1}^{M} P(\boldsymbol{X}-\boldsymbol{v}_j,t)a_j(\boldsymbol{X}-\boldsymbol{v}_j) - P(\boldsymbol{X},t)a_j(\boldsymbol{X})\,. \tag{3.3}$$

3.2.1 Example: A Simple Synthesis-Degradation Model

Let us consider a model that tracks the number of molecules m (say tokens of an mRNA sequence) produced at some rate α and degraded at a rate βm. We denote by $P(m, t)$ the probability density of finding m mRNA molecules at a time t. In a time interval Δt of duration short enough that only one of three possibilities occur: one, a mRNA molecule is produced by reaction 1 at a rate α; two, an mRNA molecule is degraded by reaction 2 at a rate β; three, the number is unchanged.

$$\emptyset \xrightarrow{\alpha} M \xrightarrow{\beta} \emptyset$$

Substituting these conditional probabilities into (3.1) and (3.2) yields

$$\begin{aligned} P(m,t+\Delta t) = {} & \underbrace{(\alpha\Delta t)P(m-1,t)}_{a_1:production:\emptyset\xrightarrow{\alpha}M} + \underbrace{(\beta(m+1)\Delta t)P(m+1,t)}_{a_2:degradation:M\xrightarrow{\beta}\emptyset} \\ & + \underbrace{(1-(\alpha+\beta m)\Delta t)P(m,t)}_{no\ reaction} \end{aligned}$$

where $a_1(m) = \alpha$, independent of m and $a_2(m) = m\beta$.

3.2.2 Computing Estimates of Variation

Gene expression levels in cells can be tracked by fluorescence microscopy, and expression levels are sorted by fluorescence intensity. Using fluorescence-activated cell sorting (FACS), distributions of protein levels can be inferred, and in isogenic populations, this provides evidence of the degree of variation in gene expression from a promoter. The observation of variability in gene expression has been further characterised in terms of 'intrinsic' and 'extrinsic' sources: variation in the expression levels of a single gene due to the stochasticity in the transcriptional process itself is called intrinsic noise, while the contribution of the fluctuations in the agents and parameters that drive transcription are collected under the term extrinsic. Operationally, two coloured fluorescent reporter genes (say red and green) driven by the *same* promoter provide visual evidence [21] of the extent to which the sources of variation of expression are global to the cell (extrinsic) or local to the promoter (intrinsic). If protein expression from both promoters occurs simultaneously, a cell in which they are produced appears yellow; else, they fluctuate between appearing red or green or yellow. Hence, the lack of correlation between red and green reporter proteins indicates the presence of intrinsic noise in the processes of transcription and translation, and correlated variation in coloured fluorescence indicates the presence of extrinsic noise.

Computing variances and correlations in molecular numbers is key to establishing degrees of stochasticity in gene expression from both experimental and theoretical approaches. Mathematically, the moments of a probability distribution:

$$\langle \boldsymbol{X}^m \rangle_t := \sum_{X} \boldsymbol{X}^m(t) P(\boldsymbol{X}, t) \tag{3.4}$$

yields the average behaviour of the system variables (for $m = 1$) as well as its deviation from the mean, e.g. (for $m = 2$) the covariances $\mathrm{Cov}(X_i, X_j)(t, t') := \langle (X_i - \langle X_i \rangle)_t (X_j - \langle X_j \rangle)_{t'} \rangle$. The mathematical details outlined here will involve computing variances and means in different biological contexts, starting with the dependence of these measures on promoter architectures. We shall not look at higher order correlations in this chapter.

We compute the time evolution of means (3.4) by multiplying (3.2) by $\boldsymbol{X} = \boldsymbol{X} - \boldsymbol{v}_j + \boldsymbol{v}_j$ and summing over all $\boldsymbol{X}$ to get

$$\begin{aligned} \langle X_i(t + \Delta t) \rangle =& \langle X_i(t) \rangle + \Delta t \sum\nolimits_{j=1}^{M} \langle v_{ij} a_j(\boldsymbol{X} - \boldsymbol{v}_j) \rangle \\ & + \Delta t \Big\{ \sum\nolimits_{X} \sum\nolimits_{j=1}^{M} \big(a_j(\boldsymbol{X} - \boldsymbol{v}_j) P(\boldsymbol{X} - \boldsymbol{v}_j, t) - a_j(\boldsymbol{X}) P(\boldsymbol{X}, t) \big) \Big\}. \end{aligned} \tag{3.5}$$

The last term $\{\cdot\}$in (3.5) vanishes by defining $\mathbf{Y} = \boldsymbol{X} - v$, and noting that $\sum_X f(\boldsymbol{X}) = \sum_Y f(\boldsymbol{Y})$. Similarly, by multiplying (3.3) by $(\boldsymbol{X} - \langle \boldsymbol{X} \rangle)(\boldsymbol{X}^T - \langle \boldsymbol{X}^T \rangle)$ and shifting variables in the sum over configurations $\boldsymbol{X}$, we get the following expression for the covariances:

$$\begin{aligned} \frac{dCov(X_i, X_j)(t, t)}{dt} =& \sum_{k=1}^{M} \langle (X_i - \langle X_i \rangle) v_{jk} a_k(\boldsymbol{X}) \rangle + \langle (X_j - \langle X_j \rangle) v_{ik} a_k(\boldsymbol{X}) \rangle \\ & + \langle v_{ik} v_{jk} a_k(\boldsymbol{X}) \rangle. \end{aligned} \tag{3.6}$$

These equations do not close on themselves if the dependence of the propensities on particle numbers ($a_k(\boldsymbol{X})$) is non-linear. For example, if $a_k(\boldsymbol{X})$ is quadratic in $\boldsymbol{X}$, the right hand side will have third order moments, while the left hand side has second order moments. A linearized approximation of the propensities around the steady states $\boldsymbol{X}^*$ ($\sum_j v_{ij} a_j \langle (\boldsymbol{X}^*) \rangle = 0$)):

$$a_j(\boldsymbol{X}) \approx a_j(\boldsymbol{X}^*) + \sum_i (X_i - X_i^*) \left(\frac{\partial}{\partial X_i} a_j(\boldsymbol{X}) \right)_{X = X^*} \tag{3.7}$$

yields the equation for time-evolution of the covariances $(\mathbf{C})_{ij} := \mathrm{Cov}(X_i, X_j)(t,t)$ expressed in terms of the parameters in the rate equations for average molecular counts and the noise statistics around steady states

$$\frac{d}{dt}\mathbf{C} = \mathbf{AC} + \mathbf{CA}^T + \mathbf{BB}^T, \tag{3.8}$$

with

$$A_{ik} = \frac{\partial}{\partial\langle X_k\rangle}\left(\sum_{j=1}^{M} v_{ij}a_j(\langle \boldsymbol{X}\rangle)\right)_{X=X^*} \quad \text{and}$$
$$\left(\mathbf{BB}^T\right)_{ij} = \left(\sum_k \langle v_{ik}v_{jk}a_k(\boldsymbol{X})\rangle\right)_{X=X^*}, \tag{3.9}$$

which is referred to as the fluctuation-dissipation theorem [61].

3.2.3 Example: Moments for the Synthesis-Degradation Model

For the simple model $\emptyset \xrightarrow{\alpha} M \xrightarrow{\beta} \emptyset$, the evolution of means (3.5) is

$$\frac{d\langle m\rangle}{dt} = \alpha - \beta\langle m\rangle,$$

yielding a steady-state value $\langle m\rangle^* = \alpha/\beta$. For the steady state variance, (3.6) for $\frac{d}{dt}\mathbf{C} = 0$ gives:

$$(-\beta C) + (-C\beta) + (\alpha + \beta\langle m\rangle^*) = 0 \Rightarrow C = \langle m\rangle^*,$$

which recovers the familiar result that the mean and the variance of a Poisson process are equal.

3.3 Promoter States and Intrinsic Noise

Unlike the simple example for mRNA production by a Poisson process just described, transcription is a complex multi-stage process particularly in eukaryotes, with a large number of protein complexes required for RNA polymerase to be poised for transcription initiation as well as for transcript elongation past paused states. Using simple models, [40, 41] explores the sources of noise (variance) in protein levels when a state transition to promoter occupancy precedes transcription

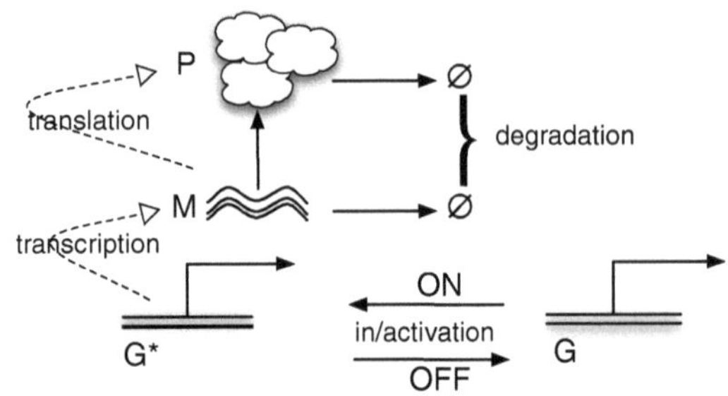

Fig. 3.1 Simple model of gene expression. The promoter switches between transcriptionally active (G^*) and inactive (G) states with rates k_{ON} and k_{OFF}. Transcription occurs from the ON state at rate $k_M{}^+$ and translation occurs at rate $k_P{}^+$. mRNA and proteins degrade at rates $k_M{}^-$ and $k_P{}^-$ respectively

followed by a translation step which produces several protein molecules from each transcript. The rates of mRNA and protein degradation determine to what extent noise in the transcription is reflected in that of the protein. Figure 3.1 summarizes the model used.

Using this simple model, we can use (3.8) and (3.9) to compute the matrix elements of the stationary equal-time covariances ($\frac{d}{dt}\mathbf{C} = 0$) using straightforward linear algebra:

$$\begin{aligned}
\sigma^2_{G^*} &= \langle G^* \rangle \frac{k_{OFF}}{k_{ON} + k_{OFF}} = \langle G^* \rangle P_{off} \\
C_{G^*M} &= \frac{k_M^+}{k_{ON} + k_{OFF} + k_M^-} \sigma^2_{G^*} \\
\sigma^2_M &= \langle M \rangle + \frac{k_M^+}{k_M^-} C_{G^*M} \\
C_{PM} &= \frac{k_P^+}{k_M^- + k_P^-} \left(\sigma^2_M + \frac{k_M^+}{k_p^- + k_{OFF} + k_{ON}} C_{G^*M} \right) \\
\sigma^2_P &= \langle P \rangle + \frac{k_P^+}{k_P^-} C_{PM}
\end{aligned} \tag{3.10}$$

What these equations capture is the intuition that in a pipeline of production processes, variability in the intermediate stages propagates to contribute to the variability of the product. In particular, the contribution of the gene activation step to the protein variablity is minimised if the promoter is always ready ($P_{on} = 1 - P_{off}$)for transcription ($k_{OFF} \ll k_{ON}$), while transcriptional and translation efficiency $k^+_{M,P}$ propagates any variability in activation.

Since all the steps involved in gene expression are stochastic, there will be variability involved which is likely to amplify in a multi-step process. Once way of reducing variability in a signal is by averaging. In this setting, the only freedom availability to the system is to absorb some of the variability in an (intermediate) input signal in the processing time for that stage. To see this more clearly from (3.10), we assume $P_{on} \approx 1$ so ${\sigma_M}^2 = \langle M \rangle$ and since $k_P^+/k_P^- = \langle P \rangle / \langle M \rangle$, we can replace two factors involving k_P^+ in the expression for the protein variance with the decay rate k_P^- and average protein and mRNA levels:

$$\frac{\sigma_P^2}{\langle P \rangle} = 1 + \frac{\langle P \rangle}{\langle M \rangle} \frac{k_P^-}{k_P^- + k_M^-} \tag{3.11}$$

This expression for a two-stage process (ignoring promoter fluctuations) emphasizes the role of time-scales of downstream processes on reducing noise levels in networks. For given mean values of mRNA and proteins, reducing k_P^- increases the number of extant proteins available to average over mRNA noise and this reduces the variability in P. This is a principle that has been examined in the context of more complex regulatory networks, where the dynamics of the network sets time scales for fluctuations to propagate [56].

It should also be noted that the k_{ON} rates of promoter activation will involve the concentrations X_{TF} of the transcription factors involved in gene regulation, each of which brings its own generated variability. This dependence is often incorporated within mathematical models as a combined average transcription rate: $M(t + \Delta t) = M(t) + f(X_{TF})\Delta t$ where $f(X_{TF})$ is described by a Michaelis-Menten or Hill form of rational polynomials. Thus, either k_{ON} or more commonly ${k_M}^+$ may be described as the rate coefficients obtained by linearization (3.7):

$$\text{production rate} \mapsto \left(\frac{\partial f}{\partial X_{TF}} \right)_{X_{TF} = X_{TF}^*} \tag{3.12}$$

where X_{TF}^* denotes steady state levels in the network. This also indicates that in more complex situations involving TF-level dependent dynamics, the regions of greatest slope in the gene regulatory function in (3.12) contributes most to the expression noise.

In this way one can envisage propagating variability in gene expression through gene networks analogously to the pipeline model of single gene expression just shown. Hence, a distinction between intrinsic and extrinsic noise [21] is made that distinguishes between the variability due to local processes (typically at a promoter, since reporter proteins are typically used experimentally to track expression levels of individual genes) and those that may arise due either to the pipeline transferring noise through the regulatory network or due to global effects (typically at a cellular level that might introduce fluctuations in the common regulatory machinery, such as polymerase or ribosome numbers).

Furthermore, in the mRNA and protein variances, the first term indicates the presence of irreducible variability stemming from the deployment of a mathematical model of a Poisson process to account for the stochastic production of molecules at a fixed rate. However, the denominators in the terms linking the variances and covariances open up the possibility of variability to be modulated, depending on the time scales over which downstream processes could average over the variability in the sample of their input components per product lifetime.

While the model described in the previous section shows that variability in protein levels can be tracked back to the rates of promoter in-activation, it is known that transcriptional regulation is often mediated by several transcription factors and slow transitions between multiple states of the promoter can limit the applicability of the results shown above. In fact, [44] have found that the limit of slow promoter transitions ($\{k_{OFF}, k_{ON}\} \ll k_M{}^+$) which increases the contribution of the promoter to expression variablity helps explain the results of experiments with native PHO5 promoters modified by upstream activating sequences (UAS) that reduce activation rates of the Pho4 transcriptional activator and thus increase noise levels. In the next subsection we will consider the possibility of multiple slow stochastic steps enabling independent control of variability that is heritable.

3.3.1 Complex Eukaryotic Promoters

A natural extension of the above model involves multiple states at the promoter. Since combinatorial transcriptional regulation is the norm, and binding site affinity could be under selectional constraints, the possibility of influencing the degree of variation around average expression levels could also be under evolutionary control. The issue of expression *divergence* between species has been examined via microarray experiments and it has been noted that promoters that contain a TATA box (all eukaryotic genomes contain TATA boxes) shows greater variability – in worms, flies, plants and animals – than those that do not [57]. The transcription preinitiation complex occupies this element, this being one of three ways in which activators influence promoter state transitions, the others being nucleosome disassembly and Pol II promoter clearance. The binding affinity of TATA box binding protein (TBP) is often a rate limiting process for transcription initiation, and a stable TBP-promoter complex can act as a scaffold for repeated binding of RNA Pol II and associated transcriptional machinery. This eliminates the need for their re-assembly after transcript formation and leads to bursts of transcription.

Here we discuss how expression levels and their variability may be subject to independent controls as suggested by the experiments and simulations of [9, 10] using the model shown in Fig. 3.2 and the theoretical work of [49, 50] which introduces a promoter state-dependent probability distribution for mRNA and protein numbers $P_s(\boldsymbol{X}, t)$ indexed by the state of the promoter s (which takes five values in the example given in the Fig. 3.2). The rates are those of association and dissociation, and the association rates are taken to be linear in each factor, the levels of which are taken to be at steady state for the computation of the variance. This linearity enables the computation of moments in the same way as for the previous case (3.10). The steady state probability of occupancy of promoter states $\pi_s = (\boldsymbol{\pi})_s$ is the stationary state of the Markov chain describing the promoter state

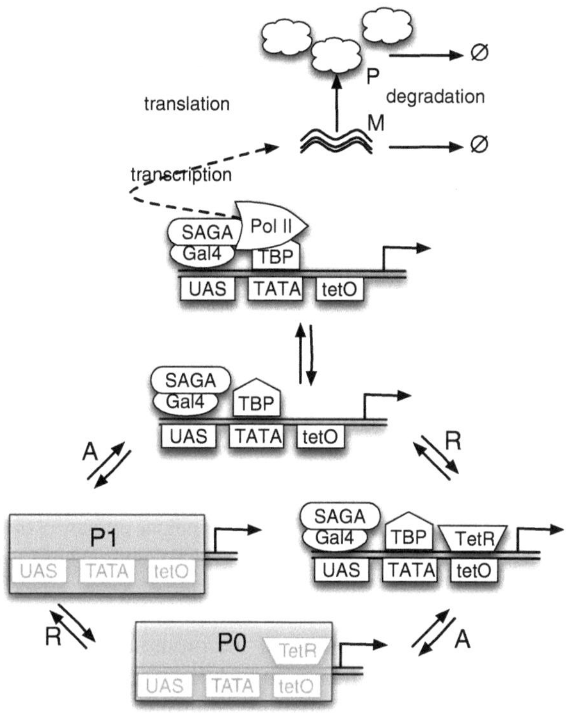

Fig. 3.2 This model extends that shown in Fig. 3.1 in containing transitions between several promoter states. The naturally occuring GAL1 promoter contains an upstream activation sequence (*UAS*) which is occupied by the Gal4 activator (in response to increasing galactose levels) and SAGA which recruits the RNA Pol II polymerase to initiate transcription with the help of the TATA-box binding protein (*TBP*). The GAL1 promoter has been engineered to contain binding sites tetO for the repressor TetR, whose binding affinity can be regulated by anhydrotetracycline (aTc)

transition matrix **K** (the matrix elements K_{ij} capture the probabilities of making transition from one promoter state j to another, i):

$$\mathbf{K} \cdot \boldsymbol{\pi} = 0. \tag{3.13}$$

Each promoter state is associated with a transcription rate $k^+_{M,s}$. For a state labelled by the number of transcripts and the internal state of the promoter, the stochastic transitions alter the state by either changing the internal state (at rates K_{ij}) or by changing the transcript level, i.e., by creating (at rate $k^+_{M,s}\ \pi_s$) or degrading (k^-_M) mRNA from internal state s. At steady state, these processes are balanced, so that

$$(\mathbf{K} \cdot \langle \mathbf{m} \rangle)_s - k^-_M \langle m \rangle_s = k^+_{M,s} \pi_s, \tag{3.14}$$

where $\langle m \rangle_s = \sum_m m p_s(m)$ is the average transcript level when the promoter is in state s. The mean and variance of mRNA levels turn out to be:

$$\langle m \rangle = \frac{1}{k^-_M} \sum_s k^+_{M,s} \pi_s \ and\ \sigma^2_M = \frac{1}{k^-_M} \sum_s k^+_{M,s} (\pi_s + \langle m \rangle_s) - (\langle m \rangle)^2, \tag{3.15}$$

with similar dependence of the protein variances on the promoter state transition matrix **K**.

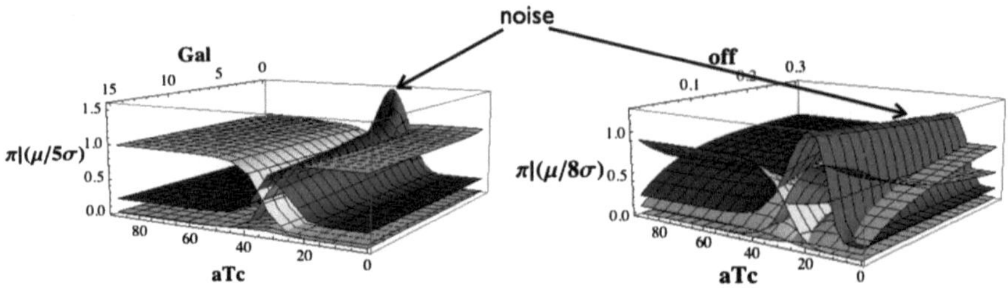

Fig. 3.3 The peaked surfaces show the noise levels (normalised as shown) for the two cases that correspond to [9] (*left*) and [10] (*right*). The other surfaces indicate the probability of promoter state occupancy as a function of aTc and Gal, to mimic the experimental controls (although not simulataneously controlled in the experiments) and as a function of aTc and a modelling parameter which captures the deactivation rate of the promoter (the A arrow to P1 in Fig. 3.2.) This off-rate changes the probability of occupation of the promoter states and thus its transcriptional efficiency. This effect is used to model the alterations to promoter sequence introduced in the experiments of [10]. Note also, that while the peak noise levels fall off as the off-rate increases. A decreased deactivation rate indicates a more stable promoter poised for elongation of the transcript, enabling bursts and hence increased noise. This is not the case away from the peak noise levels, where the increased deactivation, or weaker promoter, causes reduces mRNA production and consequenty increases noise levels

The relative occupancy of the promoter states $\boldsymbol{\pi}$ can be modulated by altering the levels of the regulator molecules (see model results in Fig. 3.3). In particular, in the experiments of [10], by controlling the levels of the inducer galactose and anhydrotetracycline (aTc) which prevents TetR binding (see Fig. 3.2) it has been shown that the noise (variance/mean) is largest in the regime when the mean levels increase with increased levels of aTc (reducing repressive effects of TetR) increasing to a peak noise level before diminishing at high values of proteins produced. In [49] it is also shown (as shown in Fig. 3.3) that the mean expression levels and its variance can be independently controlled within such a modelling scheme in a regime where the response of the expression rate to galactose saturates, thus leaving the mean expression fixed while the variance is controlled by varying aTc levels. In the context of biological networks, it is conceivable that the appropriate physiological conditions set the levels of the various transcription factors through the steady state values of the dynamical systems of these regulatory networks.

3.4 Local and Global Fluctuations: Propagation and Correlations

In what has been discussed thus far, the gene expression machinery has been characterised by parameters that set the probabilities for various events to take place. However, these very parameters could be subject to fluctuations. For instance, in the experimental settings with artificial (modified GAL1) promoters described with respect to the situation described in Fig. 3.2 the transition rates were modulated by aTc and galactose-dependent Gal4. The work of [9, 10, 49] proposed a putative mechanism for adjusting the noise of the expressed gene as a function of the externally controlled inducer and repressor levels.

In all of the above cases, the focus was on the variability of expression at a specific locus. The factors influencing the variability were captured in either the stochastic nature of the processes themselves (in the form of transcription or translation, which contributed a Poissonian variance) or in the influences of upstream fluctuations in gene regulation, such as in the dependence of the state-transitions on TF levels. The experiments of [37] indicated the importance of intrinsic fluctuations, notably the suggestion of bursting, or the translation of multiple proteins from the same transcript. Experiments investigating eukaryotic systems reported in [9, 10, 44, 46] showed up the larger contributory effects of extrinsic factors as well. While [9, 42] followed the noise in the expression of one gene down to the variability in another, [46] showed that global factors contributing variability scales with the cell cycle, or around the lifetimes of proteins.

In 'natural' networks in organisms in vivo, transcription factor levels that influence downstream expression variability are also likely to be under regulatory control and subject to variability themselves. Furthermore, since the cell is not spatially homogeneous, there could also be location dependent effects in gene expression that contribute to the observed variation. It has been known that genomic neighbourhood influences expression levels of genes (for a recent review see [17]). This has also been extended to include fluctuations in expression levels of genes, an observation that has ben used to foster speculation on evolutionary implications [4]. For instance, [6] have used yeast transgenic cells to show that integration site in a chromosome of a promoter can influence the degree of variability in gene expression. This is repeated in studies of modulation of transcriptional bursts of gene expression by integration site of the HIV-1 lentivirus [52].

3.4.1 *Spatial Embedding of Expression Dynamics: Noise and Noise Reduction*

The layout of regulatory sequences along the genome is an example of how the expression kinetics we have considered in estimating the extent of noise in expression levels accumulates and propagates. Particularly, in developmental settings, spatial inaccuracies in expression levels can lead to anomalous developmental outcomes. There are studies of the spatial heterogeneity [27] of transcription factor profiles present in *Drosophila* larvae that might present difficulties in laying out precise spatial patterns for limb formation and other developmental outcomes. It is useful to recall that the principal mechanism by which gene regulatory networks are linked together is when one protein acts as a transcription factor, turning on or off the expression of another gene. Thus far, we have pointed to the role of temporal duration or persistence of fluctuations and the time-scales over which downstream processes can delay their execution time and thus average over them. The distribution of processing in space can also suitably facilitate fluctuation reduction of regulatory outcomes at a more systemic level.

The binding of a transcription factor protein to the promoter region of a gene can only happen once the protein finds its cognate site on the DNA. This requires the proteins to find a region of the size of a nanometer within the dense and isothermal environment of the cell. Reduced accessible dimensionality to facilitate recognition events [2] is invoked to explain the fact that the association rate for the Lac repressor binding to its operator site was found [45] to be 1,000 times faster than the three-dimensional diffusion limited association rates $k \sim 4\pi D_{3D}b$ where b is the DNA base pair distance of 0. 34 nm. This speed-up is thought to be due to the intermittent switching between diffusion along the DNA and in the cell volume. Just as the rates of expression for protein production deviate from the average due to the inherent stochasticity in transcription activation and translation mechanisms, diffusional search is also a stochastic mechanism. The time taken for a protein to find its binding site is subject to fluctuations around the mean association rates as well.

A series of detailed quantitative studies was conducted on input and output levels of proteins in a transcription factor-to-expressed protein causal link. References [24, 59] studied the embryonic expression of Hunchback (Hb) regulated by the Bicoid (Bcd) transciption factor in *Drosophila*, the Hb readout being a spatial profile with a precise boundary in response to the spatial distribution of Bcd. Here the source of input noise considered was that due to Bcd diffusion in searching for the cognate binding site at the promoter driving Hb expression. The Bcd concentration level decreases exponentially from the anterior to the posterior pole of the larva and is established fairly rapidly compared to its association rate to the *hb* enhancer. However, the search for a nano-meter scale enhancer within a micro-meter sized nucleus is conducted by diffusion, which introduces a component of input noise (Bcd variability) to the output (Hb) profile. Such a component has not been considered in this chapter until now. To describe this in quantitative terms, we shall make a short digression in the next section, and introduce an additional set of theoretical tools.

3.4.2 Langevin Methods for Including Diffusion and Extrinsic Noise

The master equation (3.3) contains reaction terms in a 0-dimensional space, and must be modified to include diffusive effects. We now introduce the approximation scheme of linearlizing about steady state in (3.6) and note that the closure of moments implies that fluctuations about steady state are assumed to be Gaussian. This can be put on a more rigorous footing by viewing deviations from steady states in terms of a parameter which takes system size Ω into account [60] and letting the average behaviour be described in terms of intensive (independent of system size) variables and the fluctuations to be extensive: $\boldsymbol{X}(t) = \Omega n(t) + \sqrt{\Omega}\xi(t)$ where $\xi(t)$ are Gaussian random variables. A further digression on the relation to random walks will be helpful in motivating how modelling efforts take the time-scales of fluctuations into account.

The simplest case of a random walk adds or removes a molecule ($X(t + \Delta t) = X(t) + \Delta W(t)$, $\Delta W(t) = \pm 1$) with equal probability at every time step, and on the average a random walk takes one nowhere in particular, $\langle X(t)\rangle = \langle X(0)\rangle$. The variance in $X(t)$ grows linearly with time since

$$\begin{aligned}\sigma_X^2(t+\Delta t) &= \langle (X(t+\Delta t) - \langle X(t+\Delta t)\rangle)^2\rangle = \langle (X(t) - \langle X(t)\rangle + \Delta W(t))^2\rangle \\ &= \sigma_x^2(t) + 2\langle \Delta W(t)(X(t) - \langle X(t)\rangle)\rangle + \langle(\Delta W(t))^2\rangle.\end{aligned} \tag{3.16}$$

Since the steps are equally likely to be $W(t) = \pm 1$, the second term vanishes, and the final term is $(\frac{1}{2})(+1)^2 + (\frac{1}{2})(-1)^2 = 1$. If time τ is traversed in n such time steps, we find that the variance increases linearly with time:

$$\sigma^2(\tau) = \tau, \qquad \text{where } n\Delta t = \tau. \tag{3.17}$$

For the case of biochemical reactions, the choice of time-step Δt depends on the dynamics of the network. The Master equation (3.3) assumes that reactions occur at exponentially separated time intervals, with the exponents set by the propensities for the various reactions to take place. This requires two random numbers to be generated at each (time, reaction) paired step in simulating the evolution of the network [22]. Assuming a time step Δt that is small enough that the propensities do not change significantly ($a_j(\boldsymbol{X}, t) \approx a_j(\boldsymbol{X}', t + \Delta t)$), all reaction channels may be updated simultaneously in each Δt. If the time step is large enough to allow a number of reactions to occur, we can then use the scaling behaviour of the random walk – the variance in deviations from the average location in time t scaled as $\sqrt{t}$ – to account for the stochasticity. Thus, to compute the root-mean-square deviation of the general case of chemical random walks, we note that each reaction occurs with probability $a_j(\boldsymbol{X(t)})\Delta t$ and produces ν_{ij} molecules of X_i, thus contributing $\nu_{ij}^2 a_j(\boldsymbol{X}(t))\Delta t$ to the variance. Consequently, the update equations that incorporate the intrinsic stochastic fluctuations under this random walk approximation:

$$X_i(t+\Delta t) = X_i(t) + \sum_{j=1}^{M} \nu_{ij}\left\{\Delta t a_j(\boldsymbol{X}) + \sqrt{\Delta t a_j(\boldsymbol{X}(t))}\Delta W_j(t)\right\} \tag{3.18}$$

and the successive steps $\Delta W_j(t)$ of the random walk as well as the random increments from different reactions are uncorrelated:

$$\langle \Delta W_j(t)\Delta W_j(t+\Delta t)\rangle = 0 = \langle \Delta W_j(t)\Delta W_k(t)\rangle. \tag{3.19}$$

The presence of extrinsic noise introduces correlations between fluctuations. For instance, proteins A and B which are not in any way regulated by each other, are expressed at a rate that depends on the number of ribosomes, then a surge of available ribosomes makes both A and B levels to go up in concert and their correlations will disspate over a time scale set by the scale of the external noise, such as the time taken by the ribosomes to return to their steady-state values. This implies that the assumption of $\langle \Delta W_j(t) \Delta W_j(t + \Delta t) \rangle = 0$ will not be true unless Δt exceeds the correlation time of the external noise. A natural time scale for correlations to persist across the cell is set by the cell division time T_{div} [46]:

$$\langle \Delta W_{ext}(t) \Delta W_{ext}(t + \tau) \rangle = e^{-|\tau|/T_{div}} \tag{3.20}$$

It can be shown using (3.8) that for linear $a_j(\boldsymbol{X})$ in (3.18), $X_i(t)$ has a similar exponential correlation time. This implies that an auxiliary variable driven by a random walk process will exhibit the exponential noise correlations required to model extrinsic noise, and will be a suitable candidate for modelling purposes.

Working with linearized deviations $\delta X_i = (X_i - X_i^*)$ (3.7) from steady states enables one to compute correlation functions using (3.8). In [59] auxiliary variables were introduced to enable the application of (3.8) to compute noise correlations. This requires augmenting the **A** Jacobian matrix by extra rows/columns to include the influences of the extrinsic variables on the dynamical variables of interest.

The continuum representation of a random walk (3.18) is called a Langevin equation and is often used to model the fluctuation effects considered in this chapter. Noise driven reactions are modelled as

$$\dot{\boldsymbol{X}} = \mathbf{f}(\boldsymbol{X}) + \boldsymbol{\eta}_{noise}$$

where $\boldsymbol{\eta}_{noise}$ stands for either intrinsic or extrinsic noise. Intrinsic noise is of the form shown in (3.18), while each component of extrinsic noise $W_{ext}(t)$ is represented by an exponentially decaying average behaviour driven by a (continuous) random walk:

$$\Delta \dot{W}_{ext}(t) = \frac{1}{\tau} \Delta W_{ext}(t) + \Delta W(t)$$

of the form (3.19) which has the appropriate auto-correlation (3.20). In order to describe diffusion of the input transcription factors (input noise in [59]) diffusion terms $\nabla^2 \boldsymbol{X}$ are introduced:

$$\left(\frac{\partial}{\partial t} - D\nabla^2 \right) \boldsymbol{X} = \mathbf{f}(\boldsymbol{X}) + \eta(\mathbf{r}, t). \tag{3.21}$$

Both extrinsic noise and diffusion noise can be modelled easily by treating the stochastic biochemical processes in (3.3) as random walks – in real space as well as in the space of numbers of molecules. Correlation functions between fluctuations from steady states can be also be computed using Fourier methods which bring a frequency-domain intuition to characterising noise in gene circuits and are thus complementary to (3.8) and (3.6). It is also easier to incorporate diffusion noise using Fourier methods. Briefly, Fourier transforms are introduced as $\delta X(\mathbf{r},t) = \int d\omega d\mathbf{k} e^{i\omega t+\mathbf{k}\cdot\mathbf{x}} \tilde{\delta X}(\mathbf{k},\omega)$, for δX the vector of deviations from steady state levels, with additional spatial co-ordinates $\boldsymbol{r}$ with corresponding wave vectors $\boldsymbol{k}$. A time derivative multiplies $\tilde{\delta X}$ by $i\omega$ and spatial derivatives in ∇^2 by $-k^2$, which may then be solved for $\tilde{\delta X}(\mathbf{k},\omega)$ very simply using linear algebra and then for the fluctuations in real space and time by inverse Fourier transform.

Instead of going through the technicalities which are described in [59], we present an intuitive argument based on [7, 8, 35], using our understanding of random walks. If the step size in the random walks described above is $\pm\,\ell$ instead of $\pm\,1$, the variance $\sigma^2(\tau) = 2D\tau$ where $D = \ell^2/(2\Delta t)$. The accurate response of the *hb* enhancer is determined by its ability to 'count' the Bcd molecules reliably, and the Bcd molecule has to find the *hb* enhancer before it degrades in time $\tau_P \sim (1/k_P^-)$, during which it will have diffused a distance of $l_D \sim \sqrt{D\tau_P}$. The 'interaction volume' [35] of Bcd with the *hb* enhancer of size a is thus ${l_D}^2 a$, so the enhancer can measure around $\langle c \rangle \times a{l_D}^2$ times that number. For independently diffusing Bcd, their statistics are Poissonian $\sigma \sim \sqrt{l_D^2 a \langle c \rangle}$, and the concentration fluctuations in transcription factor input is $\delta c/\langle c \rangle \sim 1/\sqrt{\langle c \rangle a D\tau_P}$. In [24], measurements of Bcd distributions have revealed that the calculated $\delta c/\langle c \rangle$ is about six times the measured value of 10%, which is roughly an error of one nucleus width. In a simulation study, [35] propose that the mechanism that reduces the fluctuations to the observed levels is not achieved by time averaging, but spatial averaging – the Hb readout is the average of the expression of several adjacent nuclei, between which the Hb molecules diffuse. However, while further experimental results [25] on this system have corroborated this dependence of noise levels on diffusivity, the slow diffusion rates estimated indicate that as a noise reduction strategy, it has not been optimised by the system. Instead, they point to the role of autoregulatory positive feedback [32] in reducing variability. We shall discuss the role of feedback in shaping intrinsic noise next.

3.4.3 Feedback, Noise Attenuation and Amplification

Decomposing the contributions at nodes (intrinsic), edges (transmitted/extrinsic) and global (extrinsic) is the strategy used to characterise noise in networks. Synthetic networks [5] indicated a reduction of expression noise by negative feedback, whereby a protein down-regulates its own expression to reduce fluctuations. However, in (3.10) it is seen that the propagation of noise through a cascade depends on the transmission coefficient (3.12) as well as the degradation rates which specify the time scale over which the downstream components average out the fluctuations in the input. It has been shown [48, 47] that for an autoregulated gene relaxation to steady state levels occurs at different rates for activating self-regulators or repressive ones. In particular, negative feedback speeds up response times and positive feedback is slower compared to non-autoregulated genes. A sinple way to understand this is to note that deviations from steady state are enhanced for a self-activator; since at steady state production and degradation rates are balanced, this may also be viewed as a reduction in the degradation rate implying a slower relaxation to steady state. The reverse is true for auto-repressor loops. This suggests that there could be a tradeoff in the noise propagation components coming from transmission (3.12) and time averaging (3.11). In [26] a simulation analysis of three-gene networks with one gene feeding into two others which were arbitrarily inter-regulated was performed using the same framework as in (3.8). The entries in the Jacobian matrix **A** now included self-regulation terms. A simulation analysis [26] demonstrated that negative feedback attentuated noise at the cost of signal sensitivity. The reason for this was indeed due to the extended averaging time of the input due to the slower response by the positive feedback circuit.

Suppression of fluctuations by negative feedback as a control strategy requires a molecular species to sense deviations from average behaviour to perform its regulatory function. It has recently been shown [28] that this molecular sensor must be created in larger numbers than the output being sensed in order to achieve any reduction. It turns out that the standard deviation divided by the mean (a measure of noise) of the output only diminishes as the fourth root of the sensing species. Halving the standard deviation relative to the mean of the output thus requires a 16-fold increase in the amount of regulatory, sensing molecule produced, indicating that there is a severe metabolic cost to noise suppressing mechanisms. While this was a specific example of reduction in intrinsic noise reduction, metabolic cost has also been identified as a source of variation in global (extrinsic) transcription rates in human cultured cells, where the availability of diffusible ATP which is dependent on mitochondrial mass has been correlated with transcription rate variability [16].

The role of different network circuits carrying different noise signatures should be complemented with deterministic analysis of the same networks, as there is an interplay between stochastic dynamics that generates heterogeneity and deterministic dynamics that propagates information about deviations from the average behaviour and often lays out the range of possibilities that the stochastic nature of biochemical dynamics explores. We will see how positive feedback may be implemented via positive and negative interactions, i.e., via activators and repressors to generate macroscopically distinguished subtypes in isogenic populations. Before moving on to discuss the temporal characteristics of time-dependent correlations between fluctuations in activator or repressor interactions, we end this subsection with a note on the cost of noise attenuation.

3.4.4 Dynamic Noise Correlations: Activators and Repressors

As exemplified by the previous analysis, temporal scales that are set by network dynamics have significance in relaying information between nodes in a regulatory network by dampening fluctuations. In this context, the observation of fluctuations in time can provide further insights into system function. It is customary to probe regulatory dynamics by intervention – by mutating a gene, tampering with the binding activity of a transcription factor, etc. and observing the consequences.

The observation of fluctuations in gene expression opens up the possibility of conducting "natural" experiments wherein fluctuations substitute for external perturbations. The system recovers by dissipative effects to the state of homestasis which is determined by the dynamics of the network. For instance, if the system is in steady state, intrinsic noise might generate a protein burst which will be dissipated by degrading the excess proteins, or by upregulating a repressor in the same pathway. This can only be achieved via the regulatory activity of the network, hence the observation of correlated fluctuations between pairs of proteins is likely to indicate the presence of regulatory activity between them [51]. This is the statement of Onsager's regression hypothesis [36], and this reasoning underpinned Einstein's treatment of Brownian motion [20]. However, the presence of noise extrinsic to the activity of a pair of genes could cause correlated fluctuations between them and mask the observation of the intrinsic dissipative dynamics.

In [19], noise correlations between expression from regulated (via an activation link) and unregulated genes was measured to demonstrate that the activity of the regulatory interaction could be discerened from the measurement of temporal correlations, thus effectively separating out the intrinsic component that bears the signature of the regulatory interaction. In particular, the distinct signatures of extrinsic and intrinsic noise were computed in simple networks elucidating how they might help distinguish the circuit of origin from the noise correlations observed. These computations were performed using the extrinsic noise model with exponentially decaying correlations (3.20) introduced via an auxiliary variable with Langevin dynamics, and solved for the correlations using Fourier transforms. The possibility of using correlation properties of fluctuations in order to deduce regulatory behaviour has also been taken up in [15, 39, 62]. Since the propagation of fluctuations in a network occurs via a sequence of processes, each with an inherent time delay, temporal correlations in simple networks display non-monotonic behaviour in time, with peaks and troughs which characterize the response times of the network [19] with activating or repressing regulatory links. Time dependent correlations of fluctuations depend not only on the eigenvalues of the Jacobian matrix $\mathbf{A}$ in (3.9) but also on its eigenvectors [39] which are not orthogonal and whose contributions to the correlations dissipate at different rates given by their eigenvalues. The equal-time correlations, autocorrelation functions and the peaking times and peak values of the two-point correlation functions have all been proposed as features by which the regulatory network may be identified.

3.5 Noise in Regulatory Networks: Multistability and Transients

The observation and characterization of fluctuations in gene expression is important for a detailed understanding of the biology of transcriptional processes. However, regulatory networks are dynamical units that play functional roles in organisms. While phenotypic diversity has been observed and studied extensibly, a detailed picture of the networks underpinning this behaviours has only been put together in a few viral and bacterial settings, and only with synthetic biology has the detailed study of artificial networks in experimentally controlled settings been possible. Thus, the insights of the previous sections remain crucial in more complex *in vivo* studies of multicellular organisms. In gene networks, the role of noise is often dissected with reference to the contributions of the variability in the components. The average evolution and states of expressed proteins is set by the deterministic dynamics of the means (3.5), over which fluctuations smear out the trajectories of the system. Stochastic effects in networks have mostly been considered to be of importance in 'decision making' – in choosing one out of a possible number of states that are permissible given the organization of the regulatory architecture. While the setting of initial conditions, via the carefully regulated expression of genes is not ruled out, the observation of noise in expression systems requires a more through investigation of what drives decision making in development. We shall look at two principal mechanisms for state-switching behaviour. In one scenario, bistability in the underlying deterministic dynamics of the average levels of proteins is explored by stochastic effects. In the other, noise driven transients make excursions in excitable systems.

3.5.1 Multi-modality of Gene Expression in Isogenic Populations

The phenotypic consequences of noise in gene expression is most apparent in organisms or systems where fluctuations are linked to phenotypic consequences that are not explained by noise-free dynamics. If the dynamical system describing the evolution of the average levels of expression (3.5) gives rise to multiple steady states, corresponding to observable phenotypes, then the presence of noise in the system facilitates transitions between them. In [34] it was shown that the lactose utilization subsystem in *E. coli* is bistable, switching between alternative all-or-none enzyme induction states. The underlying genetic network was revealed to contain a positive feedback loop: the permease LacY which is induced via the lac operon further catalyzes the uptake of lactose. This positive feedback of a gene onto itself through intermediate steps enables stable states of high and low expression levels, separated by a threshold which is determined from a cooperative dose-response relationship between the amount of gene expressed and the inducing agent. The population heterogeneity observed near threshold levels in [34] was one of the earliest observations of bistability. Noise in gene expression facilitates the transition from one stable state to another which cannot happen for the average levels under deterministic evolution.

Mathematically, one might have, for a reduced model of two species x and y, $\boldsymbol{X} = (x, y)$

$$\dot{\boldsymbol{X}} = \mathbf{f}(\boldsymbol{X})$$

so that the implict functions of x and y where $\dot{\boldsymbol{X}} = 0$ (the x- and y-nullclines) divide the $x - y$ plane into regions where the $(\dot{x}, \dot{y})$ pairs have different signs. The points of intersection of the nullclines denote steady states, while the magnitude and direction of the vector $\dot{X}$ determines whether the system dynamics is drawn to (or away from) the steady states, for stability. (While this is easier to describe with diagram for two species, much of the analysis extends to n-species of molecules in a network.)

These stable points might correspond to phenotypes whereas the unstable points set the threshold for switching between adjacent phenotypes. Measurements of fluorescent protein reporters are pooled to construct histograms of expression levels – broad peaks of expression are indicative of stochastic dynamics which induce deviations from the steady state values of the dynamical systems. In the cases where more than one stable steady state is allowed by the system dynamics, the estimation of variability around each fixed point (using, e.g., (3.8)), provides a measure of the likelihood of crossing the threshold for state switching.

3.5.1.1 Applications of the Concept of Noise-Induced Switching

The idea that mean levels of expression in regulatory networks define developmental and physiological states, and that noise in gene expression enable switching transitions between them has been applied in many settings. We will only give a few examples, as there are numerous excellent reviews which discuss this extensively [29]. For instance, it is believed that the several thousand odorant receptors are expressed stochastically within the olfactory system. It is also believed that what reduces unnecessary duplication of this receptive repertoire is a mutual negative feedback motif which maintains the (high, low) or (low, high) levels that maintain receptor diversity. Similar mechanisms have been suggested in the developmental stages of the visual system in *Drosophila*. The ommatidia in the compound eye contain a diverse set of photoreceptors which are arranged in layers. There appear two particular receptors – 'pale' (p) and 'yellow' (y) subtypes – within the layers R7/R8 that assume their fate in a stochastic manner. Cell-type commitment in R7 (p/y) drives R8 commitment via a mutually repressive motif which receives a signal from R7.

Mutual repression (cross-antagonism [23]) of genes is also a motif that is commonly attributed to the maintenance of differentiated identities in stem cells. The networks studied in that context typically have an interlocking set of feedback loops whose dynamical systems admit multiple stable steady states each of which correspond to or trigger cell states with discrete labels of potency and/or differentiated status. In all of these examples, the average dynamics of the regulatory network determines the structure of the steady states, which correspond to the observable phenotypes, and the role of noise is to enable transitions to take place between them. Tuning the degree of expression variability would thus alter the frequency of transitions between these states.

3.5.1.2 Observing Noise-Driven Phenotype Switching

The model organism *Bacillus subtilis* has been used to validate the claims for noise driven transitions between stable states that are determined by the dynamics of the average values of protein levels. For instance, by manipulating the transcription and translation rates such that many more transcripts are generated and fewer translated per transcript than in the wild-type, a greater degree of expression variability is achieved in the bacterium, at the same expression average levels. In so doing, the probability of making a transition from one steady state to another is enhanced [33]. In a related experiment on the same organism, larger cells were created by preventing cell septation which effectively enabled a greater degree of averaging away of fluctuations and reduced the number of transitions to the other phenotypic state [55]. This phenotype is a state of competence, in which environmental DNA is absorbed and integrated into the organism's genome when growth is suppressed in times of stress. This well-characterised system contains a key protein ComK whose high expression levels induces competence. *comK* forms an auto-regulatory positive feedback loop which is necessary for competence to emerge. ComK also represses *comS* whose protein product ComS competes with the proteolytic machinery that targets ComK, thus enhancing ComK levels. The example in Fig. 3.4 is for a system with positive feedback on each gene achieved through two repressive steps, and the simplified simulation serves as a useful way for understanding these experimental results, and as an explanatory template for more complex situations, such as in the case of differentiation in embryonic stem cells. The difference in positively regulated positive feedback loops and negatively regulated (with an even number of repressive steps such as in Fig. 3.4) lies in the steady state levels of the constitutent proteins in the network as may be expected – (high, high) and (low, low) for postively regulated and (high, low) for negatively regulated positive feedback loops.

The observations of bistability caricatured by the simple model presented here have been modelled more closely on the features driving the organism to competence in the presence of quorum sensing mechanisms that increase ComS levels and trigger the transition from exponentially growing cells (no competence) to stationary cells (which enter into competence) in a relatively short time of a few hours. In the next subsection, we present an alternative dynamical viewpoint on state transitions within the ComK-ComS network over an extended time period of days in stationary (starved) conditions, and which appeals to the notion of excitability.

3.5.2 Noise-Driven Transients in Excitable Systems

The dynamics of the regulatory network which drives competence in *B. subtilis* has also been viewed as an excitable one. The reduced dynamics (assuming very fast interaction of ComK and ComS with the proteolytic machinery, compared to gene expression reactions) is shown in Fig. 3.5: the nullclines for ComK and ComS

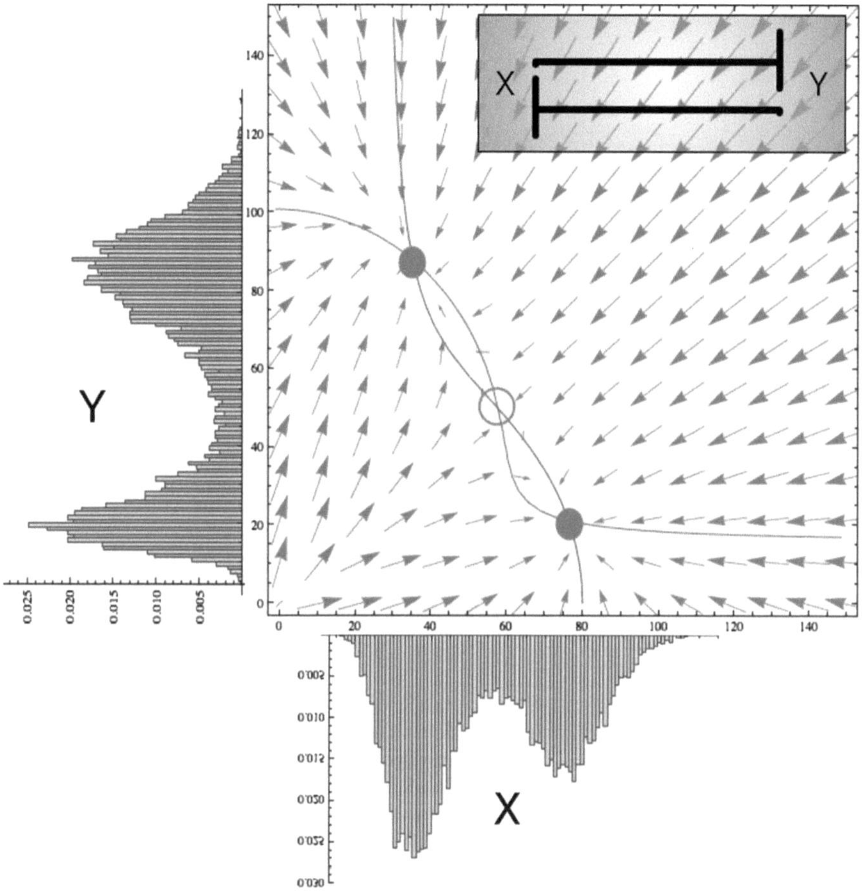

Fig. 3.4 For a two gene (denoted X and Y) model, where each represses the other as indicated in the inset, the *arrows* (vector-field) indicate the direction and magnitude of the state updates $(\dot{X}, \dot{Y})$ (see (3.5)) and the *circles* show the location of the fixed points of the dynamics ($\dot{X} = 0 = \dot{Y}$. The *open circle* fixed point is an unstable point while the convergent arrows onto the *filled circles* imply their stability. The fixed points demonstrate the (Y, Y)=(high, low) or (low, high) characteristic of mutual repression. The histograms suggests this correlation, but only a direct computation of correlations can extract that information from cell luminescence data. The nearness of the open circle (the saddle point) from the stable states is indicative of the likelihood of state switching due to noise

and a few sample trajectories from a stochastic simulation [22] are shown on a logarithmic scale. What the nullclines show are three steady states, only one of which (dark circle) is stable. The arrangement of steady states is called excitable – the rightmost point is a repellor of trajectories and the leftmost an attractor. The middle point serves as a threshold for excited trajectories – once ComK levels cross this threshold, they traverse the phase space all the way around the unstable state on the right before being drawn back to the stable state (shown by the simulation trajectories

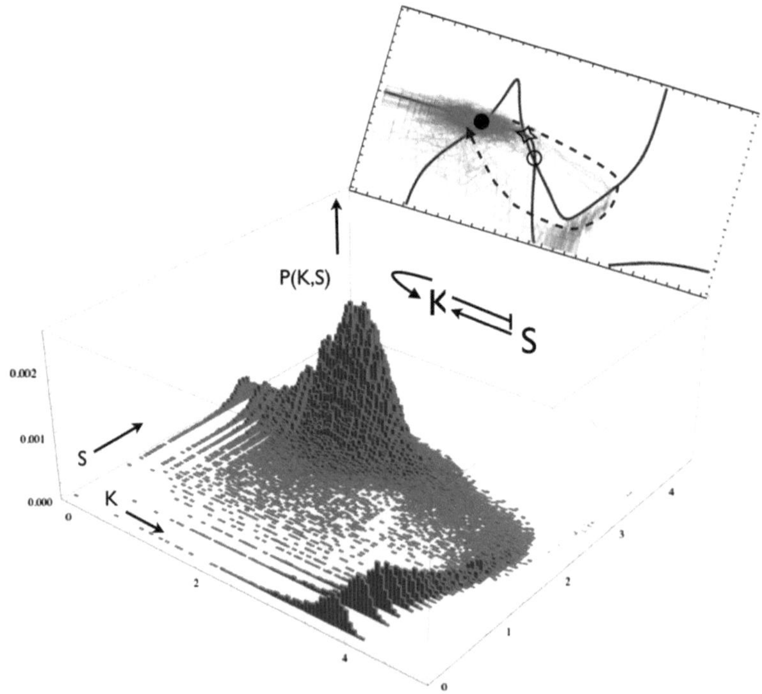

Fig. 3.5 Competence network modelled as an excitable dynamical system [55]. The excitable system is composed of two genes: *comK*, an activator which activates itself rapidly and represses the expression of another gene *comS*, while ComS reduces the degradation of ComK, in effect, acting like an activator. The excitable dynamics is illustrated by the histogram compiling the stochastic trajectories displayed in the phase diagram with the *arrow* indicating direction of traversal. The axes are logarithmically scaled. The big central peak is at the location of the stable steady state, with two additional fixed points of the dynamics (a saddle and an unstable point) ensures the existence of large trajectories in state space

which are the time series trajectories with time increasing in the direction of the dotted arrow in Fig. 3.5). The resting state of the cell in starved conditions is subject to fluctuations in expression levels, which occasionally drive the system into an excitatory trajectory and the high values of ComK expression triggers the processes leading to competence.

It has also been shown [12] that a similar excitatory network can be designed using an alternative network involving *comK* but whose excitatory duration displays reduced variability compared to the wild-type above. The authors demonstrate that there are regular environments in which this alternative circuit-driven competence states outperform the wild-type, but in more varied environments, the wild type is seen to display greater fitness. Thus while other links to evolutionary consequences of isogenic variations in gene expression have focussed on variability in levels, this study points to a time-dependent property, namely the variability in the duration of an expression burst, which leads to competence duration.

Just as the concept of noise enabled-transitions between different steady states in a multi-stable network was applied in different developmental contexts, including stem cells [13], excitability has also been proposed as a putative rationale for transitions in stem cells [30] and also to explain observed multimodality of expression distributions.

3.6 Conclusion: Controlling (for) Noise

Noise can be viewed in these studies as an enabling facility available due to the stochastic nature of the biochemical reactions that go on in cells – from search required to find appropriate partners in a chemical reaction to making the stochastic transit over the barrier to complete the reaction step. In many of the examples the diversity of regulatory responses of a cell or a promoter to internal or external cues point to the existence of stochasticity, even though its origin may not always be apparent. This has been tested in a number of studies where cell size [54], population context [53] or intracellular decision cascades [64] have been analysed to estimate the contribution of irreducible stochasticity. On a different concern, it is apparent that despite stochasticity at individual promoters, developmental trajectories in metazoans are buffered against such low-lvel noise, suggesting that regulatory interactions have been canalized [62] to eliminate such variability. In this chapter we have focused on variations in copy numbers of proteins, ignoring molecule identity. Thus the mechanisms available to explore in this chapter were of duration of processes within the regulatory network or of the spatial layout it inhabits in real organisms which enabled us to estimate the impact and propagation of fluctuations in gene expression.

References

1. Acar M, Becskei A, van Oudenaarden A (2005) Enhancement of cellular memory by reducing stochastic transitions. Nature 435:228–232
2. Adam G, Delbrück M (1968) Reduction of dimensionality in biological diffusion processes, In: Rich A, Davidson N (eds) Structural chemistry and molecular biology. Freeman, San Francisco, p 198–215
3. Balaban NQ, Merrin J, Chait R, Kowalik L, Leibler S (2004) Bacterial persistence as a phenotypic switch. Science 305:1622–1625
4. Batada NN, Hurst LD (2007) Evolution of chromosome organization driven by selection for reduced gene expression noise. Nat Genet 39:945–9
5. Becskei A, Serrano L (2000) Engineering stability in gene networks by autoregulation. Nature 405:590–3
6. Becskei A, Kaufmann BB, van Oudenaarden A (2005) Contributions of low molecule number and chromosomal positioning to stochastic gene expression. Nat Genet 37:937–944
7. Berg HC, Purcell EM (1977) Physics of chemoreception. Biophys J 20(2):193–219

8. Bialek W, Setayeshgar S (2005) Physical limits to biochemical signaling. PNAS 102(29):10040–10045
9. Blake WJ, Kaern M, Cantor CR, Collins JJ (2003) Noise in eukaryotic gene expression. Nature 422(6932):633–637
10. Blake WJ, Balázsi G, Kohanski MA, Isaacs FJ, Murphy KF, Kuang Y, Cantor CR, Walt DR, Collins JJ (2006) Phenotypic consequences of promoter-mediated transcriptional noise. Mol Cell 24(6):853–865
11. Boeger H, Griesenbeck J, Kornberg RD (2008) Nucleosome retention and the stochastic nature of promoter chromatin remodeling for transcription. Cell 133(4):716–726
12. Çağatay T, Turcotte M, Elowitz MB, Garcia-Ojalvo J, Süel GM (2009) Architecture-dependent noise discriminates functionally analogous differentiation circuits. Cell 139:512–522
13. Chickarmane V, Troein C, Nuber UA, Sauro HM, Peterson C (2006) Transcriptional dynamics of the embryonic stem cell switch. PLoS Comput Biol 2(9):e123. doi: 10.1371/journal.pcbi.0020123
14. Cohen AA, Geva-Zatorsky N, Eden E, Frenkel-Morgenstern M, Issaeva I, Sigal A, Milo R, Cohen-Saidon C, Liron Y, Kam Z, Cohen L, Danon T, Perzov N, Alon U (2008) Dynamic proteomics of individual cancer cells in response to a drug. Science 322(5907):1511–1516
15. Cox CD, McCollum JM, Allen MS, Dar RD, Simpson ML (2008) Using noise to probe and characterize gene circuits. PNAS 105(31):10809–10814
16. das Neves RP, Jones NS, Andreu L, Gupta R, Enver T, Iborra FJ (2010) Connecting variability in global transcription rate to mitochondrial variability. PLoS Biol 8(12):e1000560. doi: 10.1371/journal.pbio.1000560
17. De S, Babu MM (2010) Genomic neighbourhood and the regulation of gene expression. Curr Opin Cell Biol 22:326–333
18. Dong D, Shao X, Deng N, Zhang Z (2010) Gene expression variations are predictive for stochastic noise. Nucl Acids Res 39:403–413
19. Dunlop MJ, Cox RS III, Levine JH, Murray RM, Elowitz MB (2008) Regulatory activity revealed by dynamic correlations in gene expression noise. Nat Genet 40:1493–1498
20. Einstein A (1905) Investigations on the theory of the brownian movement. Ann der Phys 17:549–560
21. Elowitz MB, Levine AJ, Siggia ED, Swain PS (2002) Stochastic gene expression in a single cell. Science 297:1183–1186
22. Gillespie DT (1976) A general method for numerically simulating the stochastic time evolution of coupled chemical reactions. J Comput Phys 22:403–434
23. Graf T, Enver T (2009) Forcing cells to change lineages. Nature 462:587–594
24. Gregor T, Tank DW, Wieschaus EF, Bialek W (2007) Probing the limits to positional information. Cell 130:153–164
25. Holloway DM, Lopes FJP, da Fontoura Costa L, Travençolo BAN, Golyandina N, Usevich K, Spirov AV (2011) Gene expression noise in spatial patterning: *hunchback* promoter structure affects noise amplitude and distribution in *Drosophila* segmentation. PLoS Comput Biol 7(2): e1001069. doi: 10.1371/journal.pcbi.1001069
26. Hornung G, Barkai N (2008) Noise propagation and signaling sensitivity in biological networks: a role for positive feedback. PLoS Comput Biol 4(1):e8. doi: 10.1371/journal.pcbi.0040008
27. Houchmandzadeh B, Wieschaus E, Leibler S (2002) Establishment of precision and proportionality in early drosophila embryo. Nature 415:798–802
28. Lestas I, Vinnicombe G, Paulsson J (2010) Fundamental limits on the suppression of molecular fluctuations. Nature 467(7312):174–178
29. Johnston RJ, Desplan C (2010) Stochastic mechanisms of cell fate specification that yield random or robust outcomes. Annu Rev Cell Dev Biol 26(1):689–719
30. Kalmar T, Lim C, Hayward P, Muñoz-Descalzo S, Nichols J, Garcia-Ojalvo J, Martinez Arias A (2009) Regulated fluctuations in nanog expression mediate cell fate decisions in embryonic stem cells. PLoS Biol 7(7):e1000149. doi: 10.1371/journal.pbio. 1000149

31. Lehner B (2010) Genes confer similar robustness to environmental, stochastic, and genetic perturbations in yeast. PLoS ONE 5(2):e9035. doi: 10.1371/journal.pone. 0009035
32. Lopes FJP, Vieira FMC, Holloway DM, Bisch PM, Spirov AV (2008) Spatial bistability generates *hunchback* expression sharpness in the *Drosophila* embryo. PLoS Comput Biol 4(9):e1000184. doi: 10.1371/journal.pcbi.1000184
33. Maamar H, Raj A, Dubnau D (2007) Noise in gene expression determines cell fate in *Bacillus subtilis*. Science 317:526–529
34. Novick A, Weiner M (1957) Enzyme induction as an all-or-none phenomenon. PNAS 43(7):553–566
35. Okabe-Oho Y, Murakami H, Oho S, Sasai M (2009) Stable, precise, and reproducible patterning of bicoid and hunchback molecules in the early *Drosophila* embryo. PLoS Comput Biol 5(8):e1000486. doi: 10.1371/journal.pcbi.1000486
36. Onsager L (1931) Reciprocal relations in irreversible processes. i. Phys Rev 37(4):405–426. doi: 10.1103/PhysRev.37.405
37. Ozbudak EM, Thattai M, Kurtser I, Grossman AD, van Oudenaarden A (2002) Regulation of noise in the expression of a single gene. Nat Genet 31:69–73
38. Ozbudak EM, Thattai M, Lim HN, Shraiman BI, van Oudenaarden A (2004) Multistability in the lactose utilization network of *Escherichia coli*. Nature 427:737–740
39. Pakka VH, Prügel-Bennett A, Dasmahapatra S (2010) Correlated fluctuations carry signatures of gene regulatory network dynamics. J Theor Biol 266(3):343–357
40. Paulsson J (2004) Summing up the noise in gene networks. Nature 427:415–418
41. Paulsson J (2005) Models of stochastic gene expression. Phys Life Rev 2:157–175
42. Pedraza JM, van Oudenaarden A (2005) Noise propagation in gene networks. Science 307:1965–1969
43. Raj A, Rifkin S, Andersen E, van Oudenaarden A (2010) Variability in gene expression underlies incomplete penetrance. Nature 463(7283):913–918
44. Raser JM, O'Shea EK (2004) Control of stochasticity in eukaryotic gene expression. Science 304(5678):1811–1814
45. Riggs DA, Bourgeois S, Cohn M (1970) The lae represser-operator interaction: III. Kinetic studies. J Mol Biol 53(3):401–417
46. Rosenfeld N, Young JW, Alon U, Swain PS, Elowitz MB (2005) Gene regulation at the single-cell level. Science 307(5717):1962–1965
47. Rosenfeld N, Alon U (2003) Response delays and the structure of transcription networks. J Mol Biol 329(4):645–654
48. Rosenfeld N, Elowitz MB, Alon U (2002) Negative autoregulation speeds the response times of transcription networks. J Mol Biol 323(5):785–793
49. Sánchez A, Kondev J (2008) Transcriptional control of noise in gene expression. PNAS 105(13):5081–5086
50. Sanchéz A, Garcia HG, Jones D, Phillips R, Kondev J (2011) Effect of promoter architecture on the cell-to-cell variability in gene expression. PLoS Comput Biol 7(3):e1001100. doi: 10.1371/journal.pcbi.1001100
51. Sigal A, Milo R, Cohen A, Geva-Zatorsky N, Klein Y, Liron Y, Rosenfeld N, Danon T, Perzov N, Alon U (2006) Variability and memory of protein levels in human cells. Nature 444:643–646
52. Skupsky R, Burnett JC, Foley JE, Schaffer DV, Arkin AP (2010) HIV promoter integration site primarily modulates transcriptional burst size rather than frequency. PLoS Comput Biol 6(9): e1000952. doi: 10.1371/journal.pcbi.1000952
53. Snijder B, Sacher R, Ramo P, Damm E-M, Liberali P, Pelkmans L (2009) Population context determines cell-to-cell variability in endocytosis and virus infection. Nature 461:520–523
54. St-Pierre F, Endy D Determination of cell fate selection during phage lambda infection. PNAS 105(52):20705–20710
55. Süel GM, Kulkarni RP, Dworkin J, Garcia-Ojalvo J, Elowitz MB (2007) Tunability and noise dependence in differentiation dynamics. Science 315:1716–1719

56. Thattai M, van Oudenaarden A (2001) Intrinsic noise in gene regulatory networks. PNAS 98(15):8614–8619
57. Tirosh I, Barkai N (2008) Two strategies for gene regulation by promoter nucleosomes. Genome Res 18:1084–1091
58. Tirosh I, Barkai N, Verstrepen K (2009) Promoter architecture and the evolvability of gene expression. J Biol 8(11):95
59. Tkacik G, Gregor T, Bialek W (2008) The role of input noise in transcriptional regulation. PLoS One 3:e2774
60. Tsimring LS, Volfson D, Hasty J (2006) Modeling stochastically driven genetic circuits. Chaos 16:026103
61. van Kampen NG (2007) Stochastic processes in physics and chemistry, 3rd edn. Elsevier, Amsterdam, North-Holland Personal Library
62. Waddington CH (1942) Canalization of development and the inheritance of acquired characters. Nature 150:563–565
63. Warmflash A, Dinner AR (2008) Signatures of combinatorial regulation in intrinsic biological noise. PNAS 105(45):17262–17267
64. Zeng L, Skinner SO, Zong C, Sippy J, Feiss M, Golding I (2010) Decision making at a subcellular level determines the outcome of bacteriophage infection. Cell 141(4):682–691

Chapter 4
Generalised Modelling in Systems Biology

Thilo Gross

> *Open any issue of Nature and you will find a diagram illustrating the molecular interactions purported to underlie some behaviour of a living cell. The accompanying text explains how the link between molecules and behaviour is thought to be made. For the simplest connections, such stories may be convincing, but as the mechanisms become more complex, intuitive explanations become more error prone and harder to believe.*
>
> – John J. Tyson, Nature **445, 823, 2007**

Abstract A major challenge in biology is to study the functioning and failure of large networks, such as metabolic pathways, gene regulatory nets, or signalling cascades. This challenge is complicated by the nonlinearity of interactions and uncertainties in the kinetic rate laws. In a generalised model, the system under consideration is restricted to a specific structure of interactions, but the rate-laws of processes need not be specified. Thus a single generalised model can describe a whole class of plausible model systems. Despite their generality, generalised models can be investigated efficiently by methods from dynamical systems theory. They can thereby provide highly robust insights into the dynamics of the system. This chapter provides a gentle introduction to generalised modelling that is geared toward applications in systems biology.

Key words Generalised modelling • Structural-kinetic modeling • Bifurcation • Local dynamics • Mathematical modelling

---

T. Gross (✉)
University of Bristol, Merchant Venturers Building, Woodland Road, Bristol, BS81UB,
e-mail: thilo@biond.org

A. Ma'ayan and Ben D. MacArthur (eds.), *New Frontiers of Network Analysis in Systems Biology*, DOI 10.1007/978-94-007-4330-4_4,

4.1 Introduction

The lines from John Tyson, quoted above, point to an important challenge in biology [36]. What is revealed by experimental measurements is not the biology itself, but only its material basis. For understanding the biology, i.e. the emergence and functioning of life, an interpretation has to be attached to the acquired data. When simple systems are considered the process of interpretation may be so intuitive that it is often not recognized as an independent process at all – meaning seems to emerge directly from measurement. But for complex systems *intuitive explanations become more error-prone and harder to believe* [36].

Intuitive conclusions can be formalised and substantiated by conceptual mathematical models. In contrast to realistic models, which aim to replicate a system in detail, a conceptual model tries to capture a given phenomenon with a *minimal set of ingredients*. If such a model succeeds then it confirms the modeller's intuition, showing that the chosen set of ingredients is indeed sufficient to replicate the phenomenon. If the model fails it proves not only that the modeller's intuition is wrong, but by failing in a specific way provides new insights pointing to the nature of the ingredient that was overlooked. Thus, conceptual models lead to new hypothesis that can be tested experimentally or theoretically and culminate in a theory of the observed behaviour.

The mathematical modelling and analysis of complex biological systems is complicated by two factors: the large number of components that have to be taken into account and the many uncertainties that exist in the precise interaction of these components. Consider for instance cell metabolism. In all probability, future models will keep track of tens or hundreds of different metabolites in different compartments. The kinetic rate laws governing the conversion of metabolites into each other may be known for most reactions from theoretical reasoning or in vitro experiments. However, it is not clear that the same laws hold in the crowded environment of cells and under natural conditions.

Ultimately models tracking a large number of components might not be necessary, because most phenomena can possibly be explained on an emergent level of description, where the model needs to follow only a few key metabolites or aggregate variables. However, to gain the insight necessary for constructing such aggregate models, analysing models that resolve the system on a more detailed level may be crucial.

For mathematical modelling, dealing with hundreds of variables does not pose a fundamental problem. Even larger systems are routinely handled in engineering. Neither is working with uncertain interactions is itself difficult. In any analytical computation, the objects under consideration can be treated as unknowns. Nevertheless, large and uncertain systems provide a significant challenge because the numerical methods that are used for investigating large systems cannot deal with unknowns, and analytical approaches often become prohibitively difficult for all but the smallest systems.

Despite the difficulties described above, progress can still be made if we seek answers to specific questions. Consider that numerical simulation, the most common approach to model analysis, reveals whole *trajectories*, i.e. timeseries of the dynamical variables. We can say that simulation answers a very general question: "How does a system starting from some initial conditions evolve in time?". When analysing conceptual models we are often interested in more specific questions: "Is there a stable equilibrium to which the system can settle in the long run?", "Does the model explain oscillations that are observed in nature?", or "Can trajectories diverge, pointing to some flaw in the model?". These questions concern the *long-term dynamics* of the system, i.e. the behaviour observed after sufficiently long time. If these questions are answered based on numerical simulations, one considers only the last steps of long simulation runs, which offer the best approximation to the true long-term dynamics, whereas the *transients* leading up to the long term behaviour are discarded. It is therefore not surprising that methods answering questions about the long-term behaviour of the model directly (without computation of transients) are in general more efficient than simulative approaches to the same problem.

By using methods that are designed to answer specific questions we gain efficiency that can be used either to simplify analytical treatment or to speed up numerical computations. The former may allow extending the applicability of analytical approaches to larger systems. The later can be useful as it may enable running the numerical analysis more often with different parameter values, allowing the researcher to estimate the effect of uncertainties using statistical techniques. Although such sampling analysis are also commonly performed using simulation runs, methods that are geared to specific questions can be orders of magnitude faster than simulation and can therefore acquire a much larger number of samples.

In mathematics a vast array of analytical and numerical techniques for analysing specific aspects of a system's dynamics have been developed [11]. Moreover, in the applied literature different modelling approaches have been proposed that facilitate the application of specific mathematical techniques to biological applications (see [15, 26, 28, 30, 29, 35] for reviews). These include metabolic control theory [1], S-systems [27], power-law models [4] and other approaches discussed in the present volume.

In this chapter we focus on the approach of generalised modelling. Generalised models are named for their ability to capture the dynamics of systems in which the interactions between model variables is not restricted to specific functional forms. For instance a generalised model could describe a gene regulatory network in which the functional dependencies between the abundances of proteins and the rate of transcription is unspecified. Thus the generalised model can be regarded as generalisation of the class of conventional models describing the same system with specific rate laws. The approach of generalised modelling was originally developed for ecological communities [9, 5] before it was recognized as a general approach to complex dynamical systems [8].

While generalised models were at first studied only analytically, many later works used generalised models in numerical studies [31]. Interestingly, answering certain questions for the whole class of systems described by the generalised model can be easier than answering the same question for specific systems contained in this class [8, 25]. Therefore, the approach of generalised modelling can in some cases be advantageous even if the kinetic rate laws for processes in the model are known [38].

This chapter provides a gentle introduction to generalised modelling and reviews some applications of generalised models in systems biology. The central idea of generalised modelling and its mathematical background is explained in Sect. 4.2. This idea is then illustrated by a simple example in Sect. 4.3, before we discuss the application to larger models in Sect. 4.4. The chapter concludes with a brief overview of past applications in systems biology, in Sect. 4.5, and a discussion of future perspectives in Sect. 4.6.

4.2 Basic Concepts

Considering a whole class of dynamical systems, rather than a model restricted to specific functional forms, narrows the range of tools by which information can be extracted from the model. For instance we cannot simulate the dynamics of a generalised model. However, the basic tools of local dynamical systems theory can still be applied with relative ease. In this section we provide a brief introduction to these tools and sketch the basic idea behind their application to generalised models. A deeper treatment of dynamical systems theory can be found in several excellent textbooks [11, 19].

For illustration we focus on systems of N ordinary differential equations

$$\frac{\mathrm{d}}{\mathrm{dt}}x_i = f_i(x_1, \ldots, x_N), \tag{4.1}$$

where $x_1, \ldots, x_N$ are N dynamical variables and the functions f_i denote the right-hand sides of the equations. We say that a system is at a steady state, if $\mathrm{d}x_i/\mathrm{dt} = 0$ for all i. The dynamics close to a steady state, can be captured by the local linearisation given by the Jacobian matrix

$$J_{ij} = \left.\frac{\partial f_i}{\partial x_j}\right|_*, \tag{4.2}$$

where $|_*$ indicates that the derivative is evaluated at the steady state.

A steady state is locally stable if all eigenvalues of the Jacobian have negative real parts. If the right-hand side, f_i, is changed, e.g. by changing parameters on which f_i depends, then stability can be lost if at least one of the eigenvalues crosses the imaginary axis. The critical parameter sets at which such transitions occur are known as bifurcation points.

Conventional models. In *conventional modelling* one restricts the right-hand-side of the systems of equations to specific functional forms that only depend on a finite number of free parameters. In the specific set of equations that is thus obtained one can then compute the stationary solutions as a function of the parameters. Once the steady states are known, one can investigate their stability and bifurcations by explicit construction of the corresponding Jacobian matrices.

Random-matrix models. The conventional approach, described above, is routinely applied to a large variety of systems and has revealed important insights in a vast range of applications. However, it is certainly not the only way in which real world dynamical systems can be modelled. Assume for instance that too little information is available to formulate a plausible system of dynamical equations. If we unable to restrict the functions f_i to any specific functional forms then we cannot compute the steady states. Nevertheless, even without explicit knowledge of the steady states and the underlying dynamical equations, we can still express the Jacobian in the general form of Eq. 4.2. Although the entries of the matrix cannot be pinned down to specific numerical values, they are formally constants and hence can be treated as unknown parameters. If we are able to come up with plausible distributions for the values that these parameters assume in the real world then we can study the possible local dynamics in the system by random sampling.

The proposition of the previous paragraph – studying an ensemble of randomly drawn Jacobians, where assumptions on the underlying physical system only enter in the statistical distribution of matrix elements – is used in so-called *random matrix models*, which have been successfully used in important applications [13, 22]. In comparison to conventional models, random matrix models require very little information and can therefore yield robust results that are independent of many assumptions made in conventional modelling. The main drawback of random matrix models is that all information that is available on the system has to enter on the level of statistical descriptors of the matrix elements. These descriptors, the parameters of random matrix models, are in general not easily interpretable within the context of the application. It is therefore often difficult to incorporate available information into random matrix models, or, phrased differently, to restrict the ensemble of random matrices under consideration to those that can plausibly appear in the application.

Generalised models. Generalised modelling is an intermediate approach which requires less information than conventional models but allows easier integration of available information than random-matrix models. Thus, generalised models are often as easily interpretable as conventional models but come close to the elegance, robustness, and efficiency of random matrix models.

In generalised modelling we decompose the right-hand-side f_i into individual terms representing specific processes in the system, but do not restrict these terms to specific functional forms. Although we cannot compute the steady states of the system at the chosen level of generality, we can formally compute a Jacobian matrix describing all possible steady states of the system. By applying a normalization, this matrix can be expressed in terms of a number of parameters that have clear interpretations in the context of the application.

4.3 A Simple Example

For illustrating the approach of generalised modelling let us consider the example of a population of cells that increases due to differentiation of precursors and decreases due to further differentiation or cell death. This system can be modelled by a single ordinary differential equation

$$\frac{\mathrm{d}}{\mathrm{dt}}X = G(X) - L(X), \tag{4.3}$$

where X is the population size and G and L describe the gain and loss, respectively. Because we do not restrict these functions to specific functional forms, the equation above constitutes a generalised model.

Let us now consider the dynamics close to a steady state X^*. Clearly, the dynamics will depend on some properties of the unknown functions G and L at X^*. The challenge which we address is to capture these properties in a number of parameters that can be interpreted straight-forwardly in the context of the model.

For consistency we assume $X^* > 0$, $G(X^*) > 0$, and $L(X^*) > 0$. We can then define a normalized variable

$$x = \frac{X}{X^*} \tag{4.4}$$

and normalized functions

$$g(x) = \frac{G(xX^*)}{G(X^*)}, \quad l(x) = \frac{L(xX^*)}{L(X^*)}. \tag{4.5}$$

Rewriting the system yields

$$\frac{\mathrm{d}}{\mathrm{dt}}x = \frac{G(X^*)}{X^*}g(x) - \frac{L(X^*)}{X^*}l(x). \tag{4.6}$$

It can be shown [18] that the normalised system is smoothly equivalent to the original. In other words, the normalization cannot change the dynamics qualitatively. But what do we gain by normalizing the system? The normalization rescales the unknown steady state X^* such that it is mapped to a known location $x^* = 1$. Further, the unknown rates of processes in the steady state, $G(X^*)$ and $L(X^*)$, are likewise mapped to known values, $g(x^*) = l(x^*) = 1$. The price we have to pay is to introduce two new unknown prefactors, $G(X^*)/X^*$ and $L(X^*)/X^*$. Let us now study these prefactors more closely.

Because X^* is a steady state it has to satisfy the stationarity condition $G(X^*) - L(X^*) = 0$. Therefore we know that also $G(X^*)/X^* = L(X^*)/X^*$. Defining

$$\alpha = \frac{G(X^*)}{X^*} = \frac{L(X^*)}{X^*} \tag{4.7}$$

enables us to rewrite Eq. 4.6 as

$$\frac{\mathrm{d}}{\mathrm{dt}}x = \alpha(g(x) - l(x)). \tag{4.8}$$

By definition α is a constant real-valued quantity and can thus be interpreted as an unknown parameter of the model. Inspection of Eq. 4.7 reveals that this parameter has a straight-forward interpretation: It denotes the *per-capita* gain and loss rate at the steady state. In other words α is the inverse of the average time that cells spend in the population. Such parameters, defining characteristic turnover rates of the model variables are called *scale parameters* in the context of generalised modelling.

Having completed the normalization we can write the Jacobian as

$$\mathbf{J} = \alpha(g_x - l_x), \tag{4.9}$$

where

$$g_x = \left.\frac{\partial g(x)}{\partial x}\right|_1, \quad l_x = \left.\frac{\partial g(x)}{\partial x}\right|_1. \tag{4.10}$$

Again, the two quantities, g_x and l_x, can be interpreted as parameters of the model. Note that g_x and l_x are logarithmic derivatives of the original functions in the steady state. For instance

$$l_x = \left.\frac{\partial \log L(X)}{\partial \log X}\right|_*. \tag{4.11}$$

Such parameters are sometimes called *elasticities* [1]. They are used in different fields ranging from econometrics to metabolic control theory, because they can be estimated well from empirical data and provide a natural nonlinear measure for the sensitivity of the processes. If the rate of a process follows a power law, e.g. $L(X) = AX^p$ then the corresponding elasticity is the exponent p. For instance, a linear loss rate corresponds to $l_x = 1$, regardless of the slope of the linear dependency. For more complex rate laws the value of the elasticity can depend on the location of the steady state, but remains intuitively interpretable. For example an activating Hill function with exponent n, has an elasticity of n if evaluated close to zero and an elasticity close to 0 if evaluated at saturation. Inhibiting functions such as $1/X$ are charcterised by negative elasticities.

Note that the scale parameters and elasticities are defined directly at the steady state under consideration. By contrast conventional parameters often require reference to different operating points such as the half-maximum point or the growth rate at saturation, which may be unattainable in the real world system.

Using the three parameters α, g_x, and l_x we can express the Jacobian at arbitrary steady states. In this simple example the Jacobian is a 1×1-matrix and the only eigenvalue is $\lambda = \alpha(g_x - l_x)$. Stability requires that the eigenvalue is negative and therefore

$$g_x < l_x. \tag{4.12}$$

This stability condition applies to all positive steady states in the whole class of models. In particular the stability condition also holds in systems with multiple steady states. In this case the values of the generalised parameters can depend on the steady state under consideration. We remark that the generalised analysis does not prove that a positive steady state exists in a specific model. However, it is generally easy to find a specific model that has a steady state with given values of the generalised parameters [7, 34].

In summary we have shown that in all models of the form of Eq. 4.3 stability of a steady state requires that in the steady state the elasticity of the loss is greater than the elasticity of the gain. For instance, in a system with linear gain and quadratic loss the steady state is stable whereas in the case of quadratic gain and linear loss it is unstable. For the simple example considered here the reader may have guessed this result straight away. However, the same approach can be applied to large networks of interacting components.

4.4 Application to Large Systems

The central concept of generalised modelling is replacing unknown derivatives in the Jacobian by products of scale parameters and elasticities. The application of this concept to complex models is to a large extent analogous to the simple example shown above. Nevertheless, some additional difficulties appear, namely: The dynamical equations can contain more than two terms, we may have to deal with a large number of equations, and we have to explore a potentially huge parameter space.

4.4.1 Multiple Gain and Loss Terms

A relatively harmless complication is encountered, when the generalised equations contain more than one gain and one loss term. For instance

$$\frac{\mathrm{d}}{\mathrm{dt}}X = G(X) - D(X) - M(X) \tag{4.13}$$

The reader may wonder why one would need multiple terms at all, given that considering one unspecified function L covers the same range of possibilities than the sum of two unspecified functions $D + M$. The answer is interpretability. By giving the model a more detailed structure we make it easier to relate to the real world and thus facilitate the extraction of insights. If it is known that two processes act independently on a variable then it is generally advantageous to map these processes to independent terms in the generalised model. By contrast if two process are not completely independent it is generally better to describe them by a single term.

Applying the normalization procedure from the previous section to Eq. 4.13 we obtain the intermediate equation

$$\frac{\mathrm{d}}{\mathrm{dt}}x = \frac{G^*}{X^*}g(x) - \frac{D^*}{X^*}d(x) - \frac{M^*}{X^*}m(x) \tag{4.14}$$

the challenge is now to absorb the prefactors into meaningful scale parameters. Because there are now three prefactors linked by one stationarity condition $G^*/X^* = D^*/X^* + M^*/X_*$ we will obtain two independent scale parameters. It is often advantageous to define first a parameter denoting the per-capita turnover rate of the variable and then introduce further parameters that specify which proportion of the turnover originates from which process. In the present example this yields

$$\frac{\mathrm{d}}{\mathrm{dt}}x = \alpha(g(x) - \beta d(x) - \bar{\beta} m(x)), \tag{4.15}$$

where $\alpha = G^*/X^* = L^*/X^* + M^*/X^*$ is the per-capita turnover rate, $\beta = D^*/(D^* + M^*)$ denotes the fraction of the turnover occurring due to the loss D, and $\bar{\beta} = 1 - \beta$ is not an independent parameter, but the complement of β, i.e. the proportion of the turnover occurring due to the loss M.

The approach of first defining a total per-capita turnover rate and then identifying the proportions contributed by the individual processes can be applied to any number of gain and loss terms. However, care has to be take that the parameters describing the proportions add up to one for both gains and losses to ensure that the state under consideration is stationary. We note that in specific systems other ways of absorbing the prefactors into scale parameters may be advantageous. For instance in models of metabolism many conservation laws exist that reduce the number of parameters. In these systems it can be advantageous to use the flux modes of the system directly as scale parameters. An example of this approach is the analysis of the mitochondrial TCA cycle in [32].

4.4.2 Iterative Refinement

From Eq. 4.15 the Jacobian can be computed directly. However, let us compute the Jacobian in a different way that provides the opportunity to introduce a useful refinement procedure. Suppose we had just derived the Jacobian of our initial example system, Eq. 4.9. Now assume that we realize that the loss L in this model is caused by two independent processes D and M. We can now redo the normalisation for the new model, as we have done in Sect. 4.4.1, or derive the Jacobian for the extended model by *refining* the Jacobian of the initial model.

The refinement procedure starts with an equation capturing the intended refinement

$$L = D + M. \tag{4.16}$$

We normalize this equation following the same procedure that we applied for the differential equation, which yields

$$l = \beta d + \bar{\beta} m, \tag{4.17}$$

where we used $D^*/L^* = D^*/(D^* + M^*) = \beta$ and $M^*/L^* = \bar{\beta}$. By differentiating with respect to x we find

$$l_x = \beta d_x + \bar{\beta} m_x \tag{4.18}$$

Substituting this relationship into the Jacobian, Eq. 4.9 yields $\mathbf{J} = \alpha(g_x - \beta d_x - \bar{\beta} d_x)$, which is the same expression that we would have found by direct differentiation of Eq. 4.15.

The iterative refinement procedure allows making the model more concrete and therefore more interpretable without repeating the normalization of the complete model. Even when studying large systems of equations, new ideas and insights can thereby often be integrated quickly and smoothly into generalised models. For a more complex example of the iterative refinement procedure we refer the reader to [37].

4.4.3 Extracting Information from Generalised Models

In the simple example systems studied so far, the Jacobian was a 1×1-matrix, so that we were able to read off the eigenvalue directly. However, the analytical computation of eigenvalues becomes tedious already in systems with more than two variables and is generally impossible in systems with more than four variables. This problem can be overcome in two possible ways: First, we can compute the eigenvalues of a given Jacobian numerically. Second, we can compute the bifurcation points at which stability changes by so-called direct methods that do not rely on computation of eigenvalues.

Let us first discuss the direct computation of bifurcation points. Direct methods are based on the construction of test-functions that vanish at bifurcation points. Using the method described in [12, 6] such test functions can be formulated straight forwardly. The analytical expressions that are obtained in this way directly reveal the critical parameter values at which the stability of steady states is lost. Further, the test functions distinguish between different types of bifurcations corresponding to different types of instabilities, such as the onset of oscillations (*Hopf bifurcation*) or disappearance of the steady state (*saddle-node bifurcation*). Thereby, the computation of

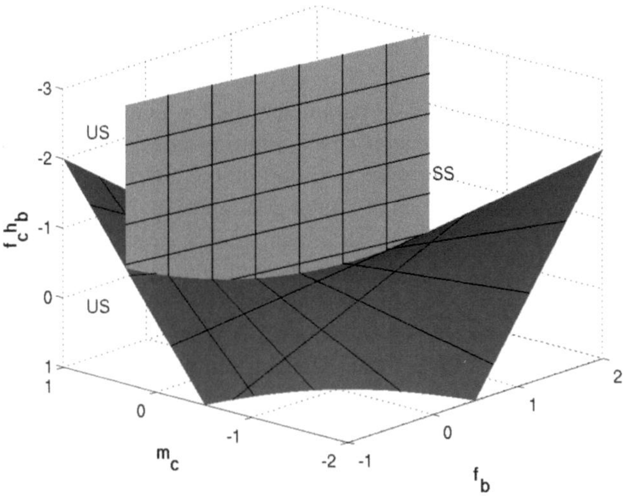

Fig. 4.1 Three-parameter bifurcation diagram of two-variable models of bone remodelling [39]. The stability of steady states in this class of models depends only on the elasticity of the net growth rate of osteoclasts with respect to osteoclast abundance, m_c, the elasticity of osteoblast reproduction with respect to osteoblast abundance, f_b, and the product of the elasticities of the growth rate of osteoclasts and osteoblasts with respect to the abundance of the other cell type, f_c and h_b, respectively. Every point in the three dimensional volume corresponds to a family of steady states characterised by the respective parameter values. Regions of stable local dynamics (*SS*) are divided from unstable states (*US*) by surfaces of bifurcation points, which correspond to different types of instabilities (*dark grey*=saddle-node bifurcation, *light grey*=Hopf bifurcation). For small generalised models such three-parameter diagrams can provide an intuitive picture of the parameter space. Larger models are better explored with statistical methods based on numerical random sampling of the parameter space

bifurcations reveals the dynamics that can be expected immediately after the stability of the steady state has been lost. Finally, the analysis can detect certain types of bifurcations (bifurcations of higher codimension) that provide some insights into the global dynamics of the system. These bifurcation can for instance indicate parameter regions where chaotic dynamics can be observed [7, 19, 38].

Using the method cited above test functions for local bifurcations can, at least in principle, be constructed for systems of any size. However, the analytical expressions that are obtained become too complicated to interpret directly for systems of more than approximately six variables. In the larger systems one can still use the test functions to visualise bifurcations in bifurcation diagrams. The major drawback of visualisation compared to direct inspection of the equations is that only a very limited number of parameters can be plotted in any one bifurcation diagram. Ideally, one uses a three-parameter diagrams (Fig. 4.1), which can be created using the algorithm described in [33].

In the three-dimensional space of a three-parameter bifurcation diagram every single point corresponds to a steady state with the respective parameter values. The bifurcation points form surfaces in the space that divide the parameter volume into regions of different stability. They thereby allow the researcher to see which combinations of parameter values lead to stable steady states and which cause instability.

Already in systems with only six parameters a three-parameter diagram can only show a small slice of the parameter space. Plotting three-parameter diagrams can still be valuable as they can provide intuition on the shape of stability boundaries and efficiently reveal bifurcations of higher codimension. Nevertheless, an exhaustive analysis of a system with many parameters cannot be based on the inspection of bifurcation diagrams alone, but must be supplemented by other techniques that use information from the whole parameter space accessible by the system.

For studying a systems with many parameters we typically have to resort to the numerical computation of eigenvalues. Because numerical eigensolvers can only work on matrices in which all parameters have been fixed to specific values, we explore the parameter space by random sampling. For this purpose we first create a large ensemble of random parameter sets. For each set we then substitute the respective parameter values into the Jacobian and numerically compute the eigenvalue with the largest real part. Depending on the sign of the eigenvalue we assign a stability of 0 (unstable) or 1 (stable) to the parameter set under consideration. Thereby, we obtain a large database of parameter sets and the stability of the steady states they describe. We can then analyse this database by statistical methods aiming to discover the features that characterise stable states.

The simplest approach for analysing the sample database is to correlate the parameter values with the stability value [31, 38]. We extract for example the value of the first parameter and the stability value from all parameter sets. Then we compute the Pearson correlation coefficient between the parameter values and the corresponding stability values. The resulting value is positive if increasing the parameter promotes stability and it is negative if a high value of the parameter hinders stability. Computing the correlation coefficients for every parameter reveals their respective effect on the stability of the system and provides some indication of their relative importance.

Once important parameters have been identified, the database of random parameter sets can be used for more detailed analyses. For instance we can create histograms that show the probability of randomly drawing a stable steady state over the value of one of the parameters. Although such diagrams show only the effect of one parameter, they contain more information than 1-parameter bifurcation diagrams, because the other parameters are not fixed to specific values. In some applications other techniques have been used to extract additional information or answer specific questions. These include the computation of mutual information between parameters and stability, the analysis of eigenvalues and eigenvectors of unstable states [38], two-parameter histograms [10], and bifurcation diagrams generated from computed eigenvalues [31].

For the computation of the largest eigenvalue, highly efficient algorithms are readily available in common software packages. Therefore sufficiently large databases for the correlation and histogram analysis can be created in reasonable computational time (often seconds to minutes) and with very little manual effort.

4.5 Applications in Systems Biology

In some applications even the analysis of very simple generalised models can reveal new insights. One such application is bone remodelling. Throughout an individual's life bone is constantly remodelled by the interplay of two cell types. Old bone is resorbed by specialised cells called *osteoclasts*, while new bone is formed by *osteoblasts*. For remaining in a physiological state it is essential that the number of osteoclasts and osteoblasts remain in balance, but in order to be responsive to changing conditions the abundances of cells must be sensitive to external stimuli.

For understanding the implications of the partially contradictory demands of stability and responsiveness it is useful to study simple non-spatial models of the regulatory interactions controlling the abundances of the two cell types. The simplest such models contain only two variables, which describe the abundance of osteoclasts and osteoblasts, respectively [17]. In some models also responding osteoblasts, a class of precursor cells, are modelled explicitly [20, 23].

Generalised models of two and three-variable systems for bone remodelling were analysed in [39]. The analysis showed that two-variable models are unlikely to have stable steady states in the experimentally observed parameter ranges, which is a potential shortcoming of such models. Further, computation of bifurcations in the general three variable model showed that the area of parameter space that is most likely realized in vivo is very close to Hopf and saddle-node bifurcations. Operating in this range enhances responsiveness of the system, but decreases its stability against perturbations. A system operating in this parameter regime may therefore be destabilized by small variations in certain parameters. Although theoretical analysis alone cannot prove that such transitions are the cause of pathologies in patients, it is apparent that a bifurcation happening in vivo would lead to pathological dynamics. In particular, a Hopf bifurcation may lead to oscillatory rates of remodelling that are observed in certain forms of Paget's disease.

The direct computation of bifurcation surfaces, was also recently used in an insightful paper that compared continuous and boolean models of regulatory networks [2]. The comparison revealed that approximating continuous time regulation by boolean functions generally changes the dynamics of the system.

A larger generalised model, which was recently proposed, focusses on the mitogen-activated-protein-kinase (MAPK) cascade, a basic motif of cell signalling that appears in all eucaryotic cells. A major question concerning the MAPK cascade is why the cascade has evolved, apparently independently, in different organisms and why it is conserved throughout evolution. It has been previously pointed out that the MAPK cascade acts as an ultra-sensitive switch [14, 21], but can also exhibit sustained oscillations [16, 24].

For gaining general insights into these dynamics of the cascade, a generalised model was formulated and analysed in [38]. In this model numerical sampling and bifurcation analysis immediately revealed the Hopf bifurcations which mark the onset of oscillatory dynamics. Furthermore the bistability, which causes the switch-like behaviour can be inferred from the presence of a saddle-node bifurcations in the generalised model. Analysing the eigenvectors close to this bifurcation provides insights into the mechanism of the switch. Perhaps most importantly, the authors found that adding additional feedback loops known in certain organisms increases rather than decreases the region in which oscillatory dynamics is observed. The latter result suggests that oscillations in the MAPK system may have functional significance, at least in certain stages of development. However, if the MAPK system indeed functions both as a switch and as an oscillator, it is not surprising that it is well conserved throughout evolution, because mutations are likely to interfere with at least one of the two functions. The generalised analysis in [38] further revealed a parameter region in which highly irregular and apparently chaotic dynamics can be observed. Whether this parameter region can be observed in experiments remains to be seen.

A considerable fraction of the generalised models studied so far describe metabolic systems. The ancestor of this line of work is [31]. Apart from a simple model for glycolysis in yeast, the paper analysed a generalised model of the photosynthetic Calvin cycle, which remains one of the most complex models that have been studied with generalised modelling. To facilitate the dynamics of this system the authors proposed a convenient notation. Generalised models using this notation are sometimes called *structural kinetic models* after the title of the paper.

One example of generalised modelling of metabolism is the analysis of the mitochondrial TCA cycle in plants [32]. This cycle plays a central role in energy metabolism and biomass production. Among other results, a sampling analysis revealed that of all processes considered in the model the import of pyruvate into the mitochondria has the strongest destabilizing effect. This may explain why pyruvate is in some plants only imported into the mitochondria in very low amounts.

Other metabolic systems that have been studied by generalised modelling include a simple model of calcium oscillations [33], a detailed model of Glycolysis [3] and general metabolic cycles [25]. The latter study showed analytically that a certain class of one-input one-output metabolic cycles is unconditionally stable, whereas autocatalytic cycles are only stable under certain conditions.

4.6 Discussion

In the beginning of this chapter we motivated the need for alternative modelling approaches with a quote from John Tyson. This quote is from an essay entitled *Bringing cartoons to life* [36], which argues that modelling approaches are needed that can extract the dynamics of the large networks of dynamical elements that are found in biology. But can generalised modelling really address this challenge? In this discussion I argue that the answer is at least partially, yes.

One major challenge that future models have to cope with is the sheer number of variables in the systems under consideration. However, in generalised modelling the number of variables is unlikely to be a serious obstacle. At least one important technique for analysing generalised models, the numerical computation of leading eigenvalues, can be scaled to systems with thousands of variables.

A potentially more serious challenge is the huge parameter space that has to be explored. The numerically efficiency of generalised modelling means that a very high number of samples can be created. For instance a recent publication [10] considered a model containing more than 5,000 unknown parameters. This model was analysed by randomly sampling the parameter space with 10^{11} sample points. However, even 10^{11} is relatively small when compared to the size of the parameter space. Consider that, placing a sample point at each corner of the parameter space would have required $2^{5,000} \approx 10^{1,500}$ samples. Nevertheless, for many analyses performed in this paper already ensembles of 10^8 parameter sets were sufficient to obtain significant statistical results with small error bars. The reason for the fast convergence is possibly that the stability boundary is relatively flat. Its properties can therefore be estimated more efficiently than the size of the embedding space would suggest. These results indicate that also future generalised models in systems biology, containing thousands of parameters, can be understood based on as few as some billion parameters sets. We believe that this number may be reduced further if the parameter space is explored by techniques from machine learning, rather than fully-random sampling.

Another considerable challenge in future models will be the formulation of the model itself. For large systems containing hundreds of variables even generating the equations can be a demanding task. In this area generalised modelling offers a clear advantage. Formulating a generalised model requires less manual work than formulating a conventional model because the functional forms in the model do not need to be specified. The subsequent construction of the Jacobian matrix can be automated in computer algebra systems and thus requires no manual work.

The price that we have to pay for the advantages of generalised modelling is that the generalised model can only be analysed with a limited set of tools. Generalised models can reveal the parameters that have a strong influence on the stability of steady states. They can be used to determine the threshold parameter values at which the stability of steady states are lost. Furthermore, they can determine the nature of the instability that caused the destabilization and reveal the dynamics that is observed immediately after the destabilization occurs. Some other information, such as the presence of bistability and in special cases chaotic dynamics, is only revealed indirectly in generalised models. Generalised modelling is therefore useful mainly for analysing local dynamics close to steady states. By contrast, if the modeller seeks to explain some transient or highly non-stationary behaviour of the system, generalised models are unlikely to be the right tool.

For studying a complex biological system, generalised models will probably need to be backed up by other modelling approaches including conventional models. In this context generalised modelling will most likely be used as a high-throughput pre-screening tool that is applied before conventional modelling is attempted. Used in this way, the generalised model can rapidly narrow down the ranges of parameters and functional forms that can possibly describe the phenomenon at hand. Using the iterative refinement procedure, described above, specific details can be integrated successively into the generalised model. Thereby, the ground is prepared for the formulation of a conventional model that in comparison reveals more detailed insights in a narrower range of parameters and functions.

References

1. Fell DA, Sauro HM (1985) Metabolic control and its analysis. Eur J Biochem 148:555–561
2. Gehrmann E, Drossel B (2010) Boolean versus continuous dynamics on simple two-gene modules. Phy Rev E 82:046120–046129
3. Gehrmann E, Gläßer C, Jin Y, Sendhoff B, Drossel B, Hamacher K (2011) Robustness of glycolysis in yeast to internal and external noise. Phys Rev E 84:021913
4. González JV, Balsa-Canto E, Wellstead P et al (2007) Power-law models of signal transduction pathways. Cell Signal 19:1531–1541
5. Gross T (2001) Population dynamics. Der Andere Verlag, Tönning
6. Gross T, Feudel U (2004) Analytical search for bifurcation surfaces in parameter space. Phys D 195:292–302
7. Gross T, Feudel U (2005) Long food chains are in general chaotic. Oikos 109:135–155
8. Gross T, Feudel U (2006) Generalized models as an universal approach to the analysis of nonlinear dynamical systems. Phys Rev E 73:016205–016214
9. Gross T, Ebenhöh W, Feudel U (2004) Enrichment and foodchain stability. J Theor Biol 227:349–358
10. Gross T, Rudolf L, Levin SA et al (2009) Generalized models reveal stabilizing factors in food webs. Science 325:747–750
11. Guckenheimer J, Holmes P (1983) Nonlinear oscillations, dynamical systems and bifurcations of vector fields. Springer, Heidelberg
12. Guckenheimer J, Myers M (1996) Computing Hopf bifurcations II. SIAM J Sci Comput 17:1275–1301
13. Jamshidi N, Palsson BØ(2008) Formulating genome-scale kinetic models in the post-genome era. Mol Syst Biol 4:1–10
14. Huang C, Ferrell J (1996) Ultrasensitivity in the mitogenactivated protein kinase cascade. PNAS 93:10078–10083
15. Guhr T, Müller–Groeling A, Weidenmüller HA(1998) Random-matrix theories in quantum physics. Phys Rep 299:189–425
16. Kholodenko B (2000) Negative feedback and ultrasensitivity can bring about oscillations in the mitogen-activated protein kinase cascades. Eur J Biochem 267:1583–1588
17. Komarova S, Smith R, Dixon S et al (2003) Mathematical model predicts a critical role for osteoclast autocrine regulation in the control of bone remodeling. Bone 33:206–215
18. Kuehn C, Siegmund S, Gross T (2010) On the dynamical analysis of evolution equations via generalized models. arXiv:1012.4340
19. Kuznetsov Yu A (2004) Elements of applied bifurcation theory. Springer, Heidelberg
20. Lemaire V, Tobin F, Greller L et al (2004) Modeling the interactions between osteoblast and osteoclast activities in bone remodeling. J Theor Biol 229:293–309
21. Markevich N, Hoek J, Kholodenko B (2004) Signaling switches and bistability arising from multisite phosphorylation in protein kinase cascades. J Cell Biol 164:353–359
22. May RM (1972) Will a large complex system be stable? Nature 238:413–414
23. Pivonka P, Zimak J, Smith D W et al (2008) Model structure and control of bone remodeling. Bone 43:249–263
24. Qiao L, Nachbar R, Kevrekidis IG et al (2007) Bistability and oscillations in the Huang-Ferrell model of MAPK signaling. PLoS Comput Biol 3:1819–1826
25. Reznik Ed, Segrè D (2010) On the stability of metabolic cycles. J Theor Biol 266:536–549
26. Rodriguez A, Infante D (2009) Network models in the study of metabolism. Electron J Biotechnol 12:1–12
27. Savageau MA, Voit EO (1987) Recasting nonlinear differential equations as S-systems. Math Biosci 87:83–115
28. Schallau K, Junker BH (2010) Simulating plant metabolic pathways with enzyme-kinetic models. Plant Physiol 152:1763–1771
29. Steuer R (2007) Computational approaches to the topology, stability and dynamics of metabolic networks. Phytochemistry 68:16–18

30. Steuer R, Junker BH (2009) Computational models of metabolism. Adv Chem Phys 142:105–251
31. Steuer R, Gross T, Selbig J et al (2006) Structural kinetic modeling of metabolic networks. PNAS 103:11868–11873
32. Steuer R, Nunes Nesi A, Fernie AR et al (2007) From structure to dynamics of metabolic pathways. Bioinformatics 23:1378–1385
33. Stiefs D, Gross T, Steuer R et al (2008) Computation and visualization of bifurcation surfaces. Int J Bifurc Chaos 18:2191–2206
34. Stiefs D, van Voorn GAK, Kooi BW et al (2010) Food quality in producer-grazer models. Am Nat 176:367–380
35. Sweetlove LJ, Fell D, Fernie AR (2008) Getting to grips with the plant metabolic network. Biochem J 409:27–41
36. Tyson JJ (2007) Bringing cartoons to life. Nature 445:823–823
37. Yeakel JA, Stiefs D, Novak M et al (2011) Generalized modeling of ecological population dynamics. Theoretical Ecology 4(2): 179–194.
38. Zumsande M, Gross T (2010) Bifurcations and chaos in the MAPK signalling cascade. J Theor Biol 265:481–491
39. Zumsande M, Stiefs D, Siegmund S et al (2011) General analysis for mathematical models in bone remodeling. Bone 48(4): 910–917.

Chapter 5
Relative Quantification Mass Spectrometry Using iTRAQ Isobaric Tags

Richard D. Unwin, Emma Carrick, and Anthony D. Whetton

Abstract Advances in mass spectrometry, separation technologies and availability of genome sequences for a number of species now offer opportunities for in depth analysis of proteomes. The sensitivity of new instrumentation means detection of low copy number proteins such as transcription factors is well developed. As biology and biomedicine are quantitative disciplines there remains the question of methods for relative or absolute quantification of these proteins. Mass spectrometry per se is not amenable to relative quantification within complex protein mixtures meaning novel approaches are required. Methods for relative quantification such as spectral counting are currently being investigated but chemical labelling methods such as isobaric tagging offer a more robust workflow allowing relative quantification of multiple samples in a single tandem mass spectrometry run. Here we describe an approach, isobaric tagging for relative and absolute quantification (iTRAQ), that has been applied with success to hematopoietic stem cell and embryonic stem cell research, as well as in clinical proteomics studies and has been used to highlight the role for post-translational regulation of protein levels in stem cell development. In this chapter, we will outline the principle of the iTRAQ methodology, provide examples of its use and discuss it strengths and weaknesses.

R.D. Unwin
Centre for Advanced Discovery and Experimental Therapeutics (CADET), NIHR Manchester Biomedical Research Centre, Manchester Academic Health Science Centre, Central Manchester University Hospitals NHS Foundation Trust, Oxford Road, Manchester M13 9WL, UK

Stem Cell and Leukaemia Proteomics Laboratory, School of Cancer and Enabling Sciences, Manchester Academic Health Science Centre, The University of Manchester, 27 Palatine Road, Withington, Manchester M20 3LJ, UK

E. Carrick • A.D. Whetton (✉)
Stem Cell and Leukaemia Proteomics Laboratory, School of Cancer and Enabling Sciences, Manchester Academic Health Science Centre, The University of Manchester, 27 Palatine Road, Withington, Manchester M20 3LJ, UK
e-mail: Anthony.whetton@manchester.ac.uk

A. Ma'ayan and Ben D. MacArthur (eds.), *New Frontiers of Network Analysis in Systems Biology*, DOI 10.1007/978-94-007-4330-4_5,

Keywords (up to 5, alphabetical order) iTRAQ • Mass Spectrometry • Proteomics • Relative Quantification

5.1 Introduction

The phenotype of a cell, its functional capability and physiologic role is determined by the proteins it expresses. The morphology, cellular secretions, rate of division and other expressions of cellularity are controlled by a vast interconnecting network of structural, enzymic, regulatory and other proteins. To understand the complex molecular machine that is the cell, the underlying genomic sequences provide a key. The genome sequencing programmes of the recent past have identified the coding sequences or genes which provide the blueprint for each expressed protein. In the human genome, for example, there are thought to be approximately 23,000 coding genes [1]. This relatively low number has been a matter of surprise to some, especially given that many different types of morphologically and functionally distinct cells in a complex organism will express protein from the same set of genes. The complexity defining an organism does not however simply come from this particular pallet of gene expression. Variations in how parts of the genes, exons, are combined to yield splice isoforms of proteins are also a feature that is able to give greater variation to the colours available using this genomic pallet. This provides the multicellular organism the capability to produce cells with the many distinct roles using the same basic toolbox. It has also become increasingly clear that the many thousands of components within a cell interact in a non-random fashion and these controlled interactions allow the cell to survive, grow, replicate, and repair itself. One key feature in the control of these protein interactions is post-translational modification; further chemical modification of the intact proteins such as cleavage, or addition or subtraction of specific chemical groups. These modifications can affect the protein location, stability, activity and lifespan. There are over 200 recorded forms of post-translational modification of proteins, thus any given protein can resemble an ornately decorated Christmas tree. In any given somatic cell type, around 30–50% of these post-translational modifications will be expressed [2]. Taking into account differentially spliced, cleaved and modified protein forms of each gene product, it has been estimated that from around 23,000 genes (not all of which are produced), there are over a million distinct protein populations in a typical cell type [3].

The assumption, then, that molecular biological approaches involving for example chromatin immunoprecipitation and mRNA microarray analyses can completely define the regulatory mechanisms underlying cellular development and function is therefore incorrect, or at least vastly over simplified. Our developing insight into the roles of micro RNAs and their ability to potentiate mRNA translation [4] adds to the need for assessment of cellular regulatory mechanisms at the protein level.

The ability to perform large scale systematic analyses on the protein complement of cells has only become possible recently, due to two major advances. The first is the availability of human and other genome sequences, which has allowed the prediction of protein sequences and hence enzymatically-generated peptide amino acid sequences. These predictions can then be used by algorithms that compare the masses of tryptic (or other protease-generated) peptide fragments generated within a mass spectrometer to *in silico*-derived predicted tryptic peptides and the fragments generated therefrom. The presence of a proteolytic peptide can thus be confirmed and hence the presence of its parent protein inferred, all based on genome sequence data. Plainly greater confidence of the presence of that protein is generated the more peptides derived from the parent protein are observed in the empirical experiment. Related to the huge opportunities for proteomics afforded by genome sequencing are the advances in mass spectrometric resolution/mass accuracy and sensitivity and advances in peptide separation methods such as liquid chromatography, which enables low abundance peptides and therefore proteins to be detected. Mass spectrometry is a fast moving field and many new instruments have become available which have ever greater advantages for the bioscientist.

Although there are subtle variations between different instruments, including the mode by which peptides are ionised (from solid phase by matrix assisted laser desorption ionisation – MALDI, or liquid phase by electropray ionisation – ESI) and the type of detector which is used for mass measurements (now typically an ion time of flight (ToF) component, or an ion trap for large scale peptide identifications) the basic workflow remains the same. Peptides eluting from a chromatographic separation are ionised and their mass to charge (m/z) values are measured. The instrument software then assesses these m/z values to determine a list of ions which fit certain criteria (e.g. they have >1 positive charge, their mass is within a certain range, they are above a certain signal threshold). The instrument then sequentially isolates these ions, and causes them to fragment (either by resonance excitation in an ion trap or more commonly by collision with an inert gas – collisionally induced dissociation or CID). This causes the peptide to fragment along its backbone and generates a peptide specific pattern of ions from which often the amino acid sequence can be read. Due to the two mass measurements taken – one for the intact 'precursor' ion and those for its fragments or 'product' ions this process is termed tandem mass spectrometry, or MS/MS. Clearly increased sensitivity, mass accuracy and resolution of a mass spectrometer will allow lower limits of detection. However new mass spectrometers can also now perform with relatively short duty cycles (the time taken to perform one scan, such as a tandem mass spectral analysis on a single peptide). The duty cycle is much less than a second thus infusing a peptide mixture is a lost opportunity because of the analyte dilution within a complex peptide matrix. Major advances in nanoflow separation technologies (reverse phase liquid chromatography for peptide fractionation, as an example) mean that peptides can be fractionated as they are introduced into the mass spectrometer. This means that at any given time point the mass spectrometer is presented with a less complex sample, thereby permitting ionization and analysis of peptides from low copy number proteins such as transcription factors without interference (to such a great degree) from higher abundance peptides (e.g. from common housekeeping proteins). Since peptides elute as a 'peak', the nature of that sample changes as other peptides elute from the chromatography column allowing more peptides to be assessed in the fractionated sample. The quality of this chromatographic separation is key, since not only does it simplify the sample matrix, sample concentration effects can also occur. The peak width for "delivery" of an analyte to the mass spectrometer can be as little as 10 s. As compared to 45 s delivery the analyte concentration will be 4.5 times higher, approximately. This therefore offers opportunities for detection of lower abundance peptides. These advances mean that the rapid and accurate identification of lower abundance proteins which have either been separated by gel-based methods, or by liquid chromatography, is now feasible with relatively little starting material.

While technologies progress and we gain the ability to identify more and more peptides from a biological sample, the major issues outstanding in the field are the identification of post-translational modifications using mass spectrometry (over 200 different types of modification have been identified on proteins to date) and the relative and absolute quantification of peptides and proteins from different samples. Measurement of post-translational modifications has been considered elsewhere [5]. We and others have previously reviewed the various protein relative quantification methods employed [6, 7]. Here we will focus on one specific method, isobaric tagging for relative and absolute quantification (iTRAQ).

5.2 Basis of the iTRAQ Method

iTRAQ is an MS/MS based quantification method, i.e. it requires peptides to be fragmented before they are quantified. Peptides within a sample are covalently modified with one of several iTRAQ tags, tagged peptides from several samples are pooled and are analysed simultaneously, providing data on peptide amino acid sequence, and its relative abundance between labelled samples. This principle was first demonstrated by Thompson et al. [8] who generated synthetic peptides containing a 'tandem mass tag' and showed that fragmentation of these peptides could release the tag, and that comparing levels of the tag gave relative quantitation. Shortly afterwards Ross et al. [9] described the iTRAQ molecule, which could add four such isobaric tags onto every peptide in four distinct biological samples. The iTRAQ technology has since been expanded to an 8plex version in 2007 [10], meaning that it can be used to compare up to eight conditions in a single experiment.

The iTRAQ label is comprised of a reporter group, a mass balance group and a amine reactive group (N-hydroxysuccinimide) which binds the tag to all peptides via a terminal amine moiety, and can also modify the free amine on the side chain of lysine residues. When labelled peptides are fragmented within the mass spectrometer, the tag is also fragmented, releasing the balance and reporter groups, with the reactive amine group remaining attached to the peptide. The reporter group varies between tags by a mass of 1 Da giving the reporter ions a mass of 114, 115, 116 and 117 Da in 4-plex iTRAQ. In 8-plex iTRAQ these same reporter ions are present alongside 113, 118, 119 and 121. The neutral mass balance group also changes in mass between the tags such that the overall mass of reporter plus balance is the same (tags are of equal mass, or isobaric) hence overall mass addition is

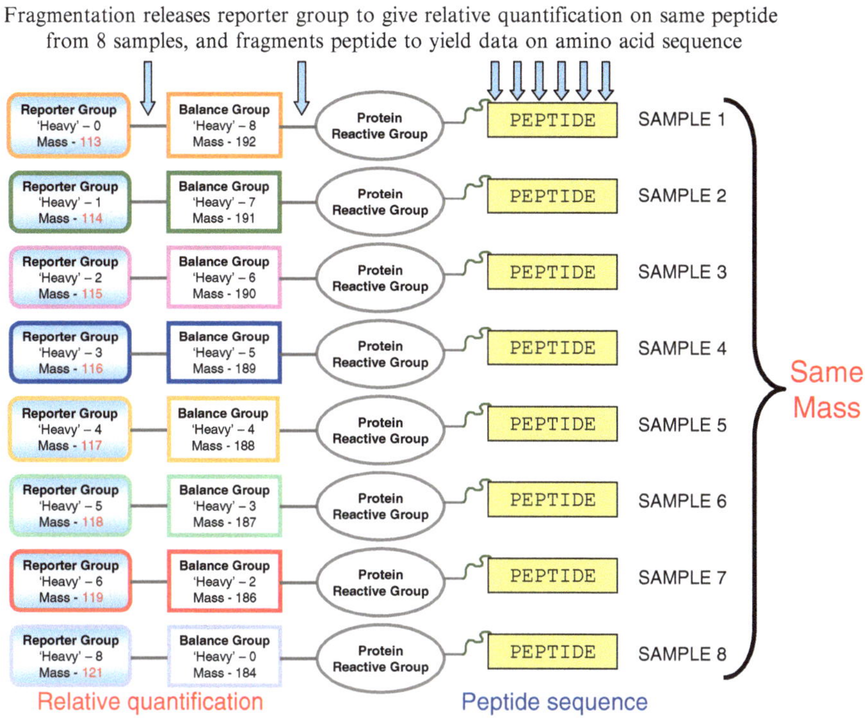

Fig. 5.1 Schematic of the iTRAQ reagent tag functional groups. An iTRAQ reagent consists of a reporter group, a balance group and a reactive group. iTRAQ tags link to the N-terminal (and lysine residues, if present) of any peptide through the reactive group, labelling all peptides in a given sample. The remainder of the tag is made of a 'reporter' group and a 'balance' group. The reporter group has identical chemical structure in each tag but has a different mass, derived from different isotopes of carbon (^{12}C vs. ^{13}C) and nitrogen (^{14}N vs. ^{15}N) reporter masses ranging from 113 to 121. There is no tag with reporter 120 Da since this would coincide with an immonium ion mass from peptides containing a phenylalanine residue. The balance group mass also varies, crucially offsetting the difference in the reporter to ensure that the overall mass of the tags (and therefore tagged peptides) remain the same. Post-labelling, samples are pooled and, since the same peptides from all eight samples has the same mass, they are indistinguishable when peptide masses are measured in a mass spectrometer. However, upon isolation and fragmentation of a specific peptide (tandem mass spectrometry, MS/MS) the tag also fragments (indicated by the *blue arrows*). Masses of fragments from the peptide (which are still the same mass from all samples, providing a moderate sensitivity enhancement since signal is summed) provide information on the amino acid sequence of the peptide. The release of the reporter ion allows comparison of reporter ion ratios in the samples, enabling determination of the ratio of that peptide between samples

constant for each of the iTRAQ labels (Fig. 5.1). In actual fact, the tags are not exactly isobaric due to the nature of the heavy isotopes used in their construction. The consequences of this are discussed in more detail later in this chapter.

The proteomics workflow for quantitation by iTRAQ usually follows a similar path (Fig. 5.2). A protein extract is digested into peptides using a site specific protease, usually trypsin which cleaves protein following each lysine (K) or arginine (R) residue, except where the following amino acid is a proline (P). Peptides from different

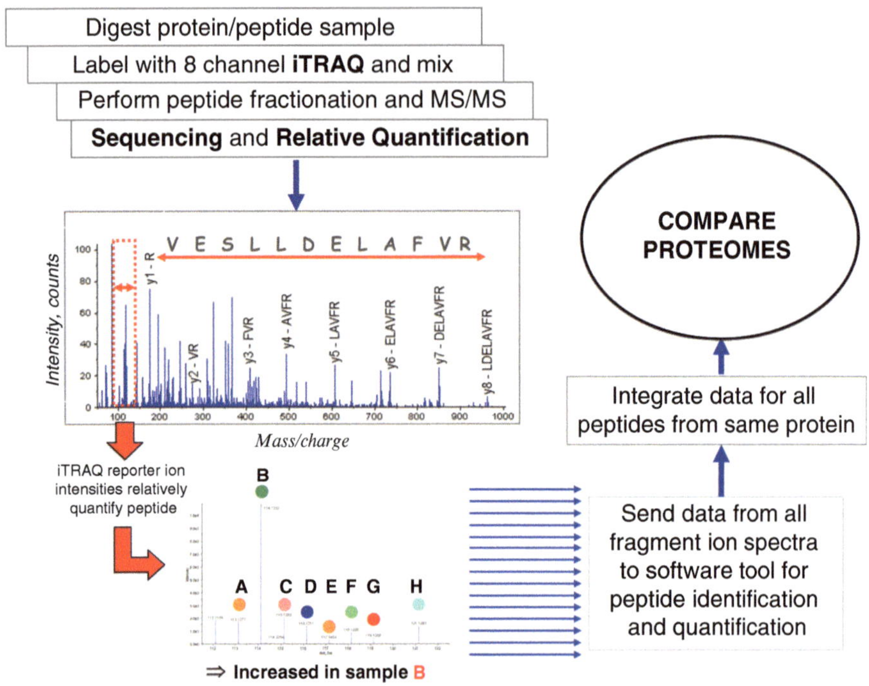

Fig. 5.2 A typical iTRAQ workflow. Proteins from up to eight different samples are isolated and digested with a specific enzyme to into shorter peptides. These peptides are then labelled with one of eight iTRAQ reagents, and samples are pooled. The pooled, labelled peptides subsequently undergo up to two rounds of chromatographic separation before being introduced into the mass spectrometer, where they are isolated according to the mass:charge ratio, and individual peptides are fragmented. This yields a fragmentation spectrum as shown in the figure, which provides information about the peptide sequence, and the release of the iTRAQ reporter ions provides information about the relative quantity of this peptide across the eight samples. A full experiment can yield tens or hundreds of thousands of such spectra, and these are analysed by dedicated computer algorithms which match the pattern of fragment ions with predicted patterns from protein sequence databases, and also extract data on the size of the iTRAQ reporters to identify and relatively quantify peptides. Once spectra have been analysed, a second set of algorithms integrates the peptide data to identify the proteins present in the sample and the relative levels between the eight samples

experimental conditions are labelled, then combined prior to any downstream enrichment or fractionation. By combining experimental conditions at this point it is assumed that any loss of peptides throughout sample preparation will be equal in all the combined samples. Pooled, labelled samples are then analysed by tandem mass spectrometry. Each fragmentation spectrum is then assessed to provide information regarding the amino acid sequence of the peptide, while the relative abundance of the peptide between the samples is reflected in the abundance of each of the reporter ions generated.

The original iTRAQ paper by Ross et al. analysed iTRAQ peptides using a MALDI-ToF instrument. Since iTRAQ was introduced in 2004 the mass spectrometers used to analyse these peptides were those with a time of flight analyser including MALDI-TOF and Q-TOF instruments. The ToF analyser was and still is predominately employed in the analysis of iTRAQ peptides due to iTRAQ experiments requiring an MS/MS spectrum with ions in the low mass region (100–200 m/z, or Thompsons, the unit for m/z). The requirement of the analysis of low mass ions in iTRAQ does not lend itself to Collision Induced Dissociation fragmentation in either an ion trap or ion trap hybrid instrument such as a LTQ-Orbitrap, since these instruments do not readily detect fragment ions of less than 1/3 of the precursor mass. Different fragmentation methods have been developed for such instruments, for example Pulsed-Q Dissociation (PQD) [11] and High energy C-trap dissociation (HCD) [12, 13] in an LTQ-Orbitrap to enable the quantification of low mass reporter ions. Electron Transfer Dissociation (ETD) [14–16] has also been used for the fragmentation of iTRAQ labelled peptides. This mode of fragmentation is useful for analysis of post-translationally modified peptides as the modification remains intact, as opposed to CID, which in the case of phosphorylation, removes the modification and can make identification of its location problematic. However when used in conjunction with iTRAQ the tags fragment differently such that the 4 plex tags generate only three unique reporter ions and the 8plex tags generate only five [14–16].

iTRAQ offers several advantages over other methods of quantification such as stable isotope labelling of peptides in culture which quantify peptides based on the relative abundance of differentially labelled intact peptides, rather than the relative abundance of tag-generated peptide fragment ions. The iTRAQ approach does not require labelling at the organismal or tissue culture level with "heavy" isotopomeric essential amino acids as the stable isotope labelling method (often termed SILAC) does. It can therefore be used for biological fluids from clinical trials, for example. Another major advantage is in time course experiments where more than three samples are to be analysed. Such experiments can be performed using SILAC [17, 18] but this requires the samples to be divided between two distinct MS runs, with a common sample in each run. However, differences in the peptides which are identified in two distinct mass spectrometry analyses means that in such experiments only a proportion of peptides identified are found in every experimental run from which data is to be amalgamated. The extended multiplex nature of iTRAQ means that a peptide is identified, and therefore quantified, in all experimental conditions simultaneously. This allows simpler experimental designs for time courses e.g. of drug treatment in cell line studies or for biomarker discovery to name but two applications. There are few if any relative quantification methods which can rival the ability of 8 plex iTRAQ to combine experimental conditions in such a way thereby minimising use of machine time. It is conceivable that label free approaches can be more effectively used when there are more than eight samples in a study. However, if performed correctly the label free approach requires extensive mass spectrometry time and very reproducible chromatography to obtain high quality data.

As a drawback, since samples are pooled fairly late in the process iTRAQ does not lend itself to complex sample processing workflows such as the isolation of a particular class of proteins or cell fraction prior to labelling. The SILAC approach, where cells can be pooled and the isolation carried out on a single pooled sample is beneficial in generating fewer errors. This kind of study can be performed with iTRAQ [19] but it is important to include technical replicates to assess the reproducibility of the sample preparation strategy and also to minimise variations introduced prior to iTRAQ labelling.

The isobaric nature of iTRAQ also potentially offers an advantage over other quantification methods in terms of sensitivity of the analysis. The signal-to-noise ratio in mass spectrometry is an important consideration especially in studies where the most interesting proteins tend to be less abundant, or in studies where sample is limited e.g. in studies of rare primary cell types. The greater the signal-to-noise ratio the greater the probability that a precursor ion will be selected for MS/MS and subsequently identified. The use of iTRAQ labelling and the pooling of samples greatly increases this signal-to-noise ratio for most peptides present in a given sample. This in turn increases the probability that a lower abundance precursor ion will be selected for MS/MS. In our experience this increase in signal-to noise has been of greater benefit when sample quantity is limited. The use of iTRAQ labelling can also improve the MS/MS spectra acquired for peptide identification due to an increase in product ion intensity from the integrated samples (and fragment ion signal-to-noise, since signal is once again summed from all eight samples) and as such increase the confidence of peptide assignment upon database searching. It is critical when performing such studies to know that the presence of the iTRAQ label on a peptide does alter its optimum fragmentation energy. iTRAQ labels will absorb collision energy increasing the chances of incomplete fragmentation and resulting in a poor MS/MS spectra and a subsequent low confidence in peptide assignment. However this issue can be overcome by optimising the collision energy used specifically for iTRAQ-labelled peptides. Interestingly, for iTRAQ-labelled phosphopeptides collision energies need to be increased further as these species require three separate events i.e. loss of the phosphate, fragmentation of the iTRAQ molecule and fragmentation of the peptide backbone. iTRAQ addition can also alter the pattern of fragment ions. For example fragments which contain an internal iTRAQ labelled lysine residue (either as a result of a missed enzymatic cleavage or the presence of a -KP- amino acid sequence) tend to retain two charges, and thus appear at a different m/z value compared to similar fragments from an unlabelled peptide.

5.3 iTRAQ Method-Based Experimental Design Is Critical for Optimal Data Acquisition

Of crucial importance to any experimental design is the requirement for appropriate comparison and controls. For all experiments, and probably more so for complex large scale screening experiments such as these, the use of biological and technical replicates is vital for complete and confident analysis of an experimental system. The technical replicate will take into consideration the ability of peptides to be analysed by mass spectrometry, as well as the reproducibility of the proteomic workflow used. Due to the nature of natural biological variation, even in identical populations of cultured cells, replicate samples are inherently different and as such replication helps determine which changes are as a result of biological intervention, rather than a natural variation.

Initial iTRAQ work up experiments should ideally consist of biological and technical replicates. Technical replicates allow the user to assess variations introduced during sample handling. This is very important, especially in more complex, multistep sample handling workflows where each step has the capability of adding a little more variation. Including a biological replicate here is useful as this allows the user to determine whether any variation in their systems is due to the nature of the sample itself, or the workflow being employed. The data generated from these initial experiments can also enable the researcher to assign minimum and maximum cut off ratios to define proteins being down or up regulated in the true experiment [20, 21]. Biological replicates also assess the consistency of any changes which are observed during the screening experiment, and only those changes observed in a majority of replicates should be taken as being reliable (Fig. 5.3).

5.4 Turning iTRAQ-Based Experimental Data into Information: Data Analysis

Once an iTRAQ experiment has been designed, prepared and analysed using mass spectrometry it is a common misconception that the result of the experiment is close at hand. However it is imperative that extensive data analysis is carried out and robust statistical methods implemented in order to assess the accuracy of all quantification methods. It is the quality assessment analysis of a mass spectrometry run that enables researchers to come to a conclusion on biological truths, if any are to be found. Analysis of these data involves a probabilistic prediction of peptide sequence based on how closely the pattern of fragment ions in any spectra match the predicted pattern of ions generated *in silico* from databases of protein (and therefore peptide) sequences. This provides putative peptide identification. The relative ratio of that peptide between samples can then be calculated based on the ratios of the iTRAQ reporter ions. Finally, since many proteins are identified and quantified based on several peptides, the data from each constituent peptide is combined to generate an overall protein ratio. Since the peptide identification software for this process is reviewed in detail elsewhere [22] we will not consider it further here. However there are a variety of tools and methods which have been developed to quantify peptides and proteins which will be discussed below.

There are several analysis packages available for iTRAQ data sets. Probably the most sophisticated is Protein Pilot from ABSciex [23]. Key features of this

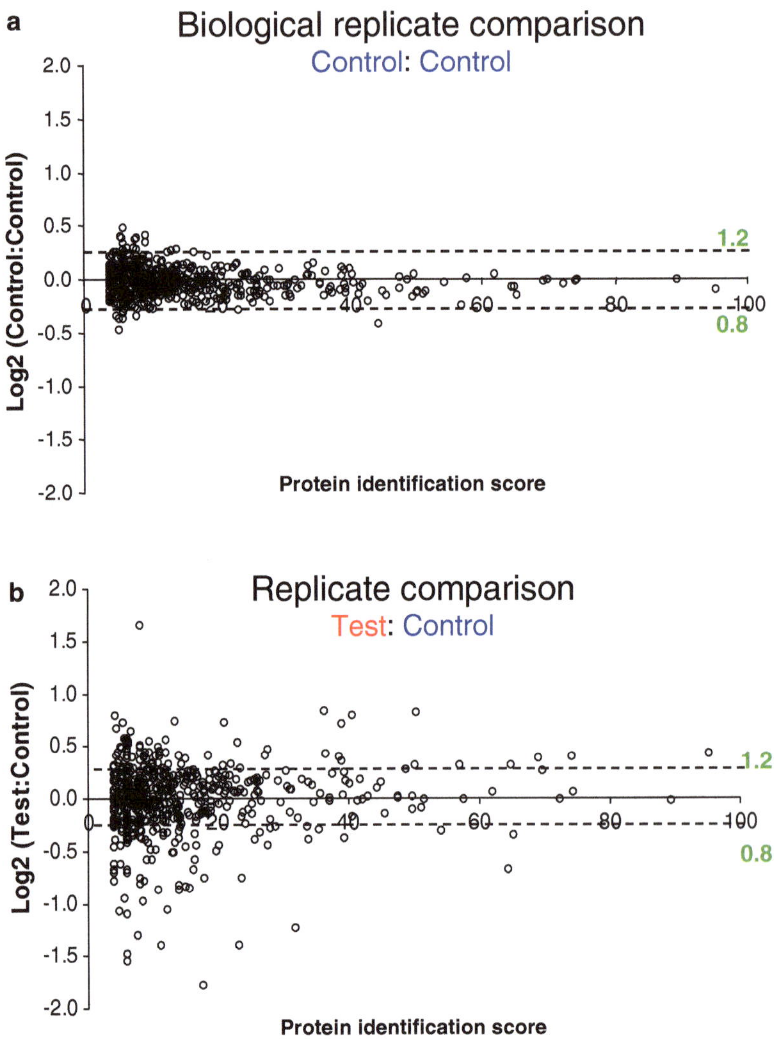

Fig. 5.3 (a) **Representative data from an internal replicate control experiment**. Here, biological replicates of the same sample have been analysed in two separate channels. This allows the technical and biological variation within the experiment to be estimated, and enables determination of ratio cut-offs outside which a protein ratio is likely to be due to a real biological effect rather than experimental variation. This plot shows the log2 ratios of the control versus control comparison against the protein confidence derived from the number of peptides. This shows that firstly protein ratios are more accurate the greater the number of peptides contribute to it, and secondly that most proteins are well quantified, with >95% of proteins lying between ratios of 1.2 and 0.83. (b) **Representative data from a test versus control experiment.** This plot shows the spread of log2 ratio data from a test versus control comparison against the protein confidence derived from the number of peptides. Comparison of this distribution with that from the control: control experiment shows the greater spread of protein ratios, with those proteins lying outside the control:control distribution being defined as changing between the test and control in this experiment

package from the point of view of quantification include 'weighting' of peptides when generating protein ratios, such that the peptides with the larger iTRAQ reporter areas (normally giving a more accurate reflection of the true ratio) contribute more to the protein ratios than a peptide with weak iTRAQ signal. ProteinPilot also has options for including or excluding peptides with different post translational modifications and generates statistics such as p-value and 95% confidence interval for each protein ratio between samples. Proteome Discoverer enables database searching and quantification analysis on Thermo instrument raw files and SpectrumMill from Agilent has tools for analysis of iTRAQ data but these currently lack the sophistication of the ProteinPilot approach.

Several other tools for iTRAQ quantification have also been made available, including iTracker [24], Multi-Q [25], iTRAQPak for R [26], iQuantitator [27], VEMS [28], TandTRAQ [29], PEAKS [30] and Census [31]. Each of these tools uses slightly different approaches to determining the peptide and protein ratios.

Once an analysis has been performed, several quality control measures should be assessed to ensure that the experiment has worked correctly, and that conclusions can safely be drawn. First of all is the extent of iTRAQ labelling: rate of labelling of lysine residues should be calculated (this is a better indication of labelling efficiency than N-terminal labelling, as N-termini can be chemically blocked, or new termini generated post labelling by peptide hydrolysis or in-source fragmentation in the MS). Data should be normalised such that the median ratio is around 1, thus correcting for slight errors in protein measurement or pipetting. Data should be checked to determine that the spread in protein ratio values is approximately normally distributed (hence the statistics which are applied are appropriate). Finally the duplicate samples in the experiment should be assessed to ensure that the majority of protein ratios fall within specific ratio limits, and then those limits can be applied to determine a cut-off for a 'significant' fold change.

5.5 Outstanding Issues with iTRAQ Methodology

The advantages of iTRAQ over other tools for protein relative quantification are generally based around the multiplex nature of the reagent, allowing 8-point experiments such as time courses or dose response curves while still enabling analysis of all samples simultaneously which results in reduced noise. The isobaric nature of the reagent and its basicity also provide a theoretical sensitivity enhancement. However there remain several criticisms and issues surrounding iTRAQ methodology.

Probably the major issue in the use of iTRAQ for protein quantification is that the protein ratios reported can be inaccurate. In effect, iTRAQ approaches underestimate the degree of fold change (so a real doubling of a protein concentration may be reported as a 1.6-fold change, for example). This observation was noted in the first iTRAQ publications where standard protein mixtures were used to test the reagent [19] and several theories have been proposed why this should be the case, and whether it can be corrected.

This underestimation may be because of the presence of co-eluting peptides of similar mass to that peptide being targeted. Despite the accuracy and resolution of the mass analyser in the MS being more than capable of distinguishing two peptides of similar (not identical) mass co-eluting, the same cannot be said for the quadropole filter which isolates the peptide prior to MS/MS. It is proposed that in some cases a proportion of the contaminant peptide is also fragmented, and iTRAQ reporters from such contaminant peptides (which are highly likely to be equal since most peptides ratios between samples/channels in an analysis do not change) contributing to a 'background' in the iTRAQ reporter channels which dilute out the real difference. This feature was demonstrated by the group of Joshua Coon (American Society for Mass Spectrometry annual conference, 2010), who noted that where a protein is identified with many peptides, it was possible to find peptides which accurately reflected ratio and some which underestimated (and hence when a protein-wide ratio was generated it was an underestimate). In a majority of cases re-inspection of the data suggested that indeed a contaminating co-eluting peptide was present and post-acquisition removal of these spectra improved the accuracy of quantification. A study by Karp et al. [32] tried to replicate this by deliberately opening up the width of the peptide isolation m/z value window and, while they saw an effect on ratio accuracy, they could not demonstrate a strong effect, although sample complexity in their data was relatively low and as such it remains a possibility that underestimation increases with sample complexity.

Ow et al. [33] have also highlighted the issue of isotopic impurities in the iTRAQ reporter region which can interfere with ratio reporting. These isotopic impurities generally manifest themselves as being an additional contribution at a mass of +1 i.e. the 113 reporter ion will contribute a small, fixed amount to the 114 channel etc. Isotopic correction factors are not supplied with the iTRAQ kits, although default values are available from the vendors in their ProteinPilot analysis software. These variations can be easily corrected for, post acquisition. In addition for phenylalanine containing peptides, which yield an immonium ion at mass 120, this ion can contribute to the 121 channel due to the presence of a naturally occurring ^{13}C isotope. Once again, this can be accounted for as the interference is a fixed proportion of the area of the 120 peak (in the region of 9.5%, this figure can be determined experimentally by analysis of unlabelled Phe-containing peptides). Some of these factors can theoretically be corrected for. Karp et al. [32] have shown that a modelling approach can be used to reduce variability and underestimation in iTRAQ datasets, as well as reducing noise and improving data precision. As the field matures, this kind of modelling approach for iTRAQ data analysis will become the norm and further enhance its value in multiplex quantitative proteomics.

Another factor which may affect the accuracy of iTRAQ quantification is the quality and specificity of the labelling reaction. As well as labelling peptide N-termini and lysine side chains, the iTRAQ reagent can also modify Cysteine, Tyrosine, Serine and Threonine residues. To achieve good quality quantification using any chemical strategy it is imperative that the labelling is complete on those residue which are modified. In the case of iTRAQ, it is impossible to get complete modification on all of the above residues, and so when designing the buffers for labelling, careful consideration must be taken to maximise labelling of N-termini and lysine residues but minimise the others. Practically, cysteine labelling is blocked by reduction and alkylation of all cysteines prior to digestion. Labelling of tyrosine is more efficient under acidic conditions, so the use of a buffer which can keep the reaction basic despite the acidic nature of the iTRAQ reagent itself (i.e. 1M Triethylammonium bicarbonate) effectively minimises modification on tyrosine. Labelling of serine and threonine residues is chemically harder to control, but tends to be low level (<5%) and usually on peptides which also contain a histidine. This is a common side reaction for NHS-ester peptide modifications [34] and is not specific to iTRAQ *per se*. This modification of serine with a nearby histidine occurs at a much slower rate than modification of lysine and peptide N-termini, and optimisation of reaction times should allow this modification to be minimised.

Another feature of the iTRAQ technology which one must be aware of is that the tags are not precisely isobaric. Due to the different use of ^{15}N and ^{13}C isotopes in the structures, the exact masses of the 4-plex iTRAQ tags are 144.1059 Da for the 114 label, 144.0995 Da for the 115 label and 144.1021 Da for the 116 and 117 labels. For the 8-plex labels, this is slightly simpler, with the 115, 118, 119 and 121 labels weighing 304.1990 Da, and the 113, 114, 116 and 117 labels weighing 304.2054 Da. This seems a very minor difference in mass and indeed most mass spectrometers operating at a resolution of 50,000 or fewer will fail to completely resolve the two species when attached to a peptide of, say 10 amino acids in length (peptide mass ~1100.5 Da, add iTRAQ label = 1404.7 +/−0.0064 Da, doubly charged = m/z 703.35 +/−0.0032). However this could affect database searching. In theory this should be a soluble problem as the masses of the iTRAQ modifications are known and hence one can still calculate the mass of the peptide with high accuracy, although as yet no current software tools of which we are aware take this into account. It has no effect on quantification since the quadropoles which filter out the ion of interest prior to fragmentation have a resolving power between 0.2 and 1 Da and as such both masses of iTRAQ modified peptide are transmitted with effectively equal efficiency.

It is worthy of note at this point that due to its applicability to all types of samples, iTRAQ is much more amenable to the analysis of standard protein mixtures and therefore such thorough characterisation in terms of how well it reflects the true (relative) quantity of protein. For SILAC methods, for example, it is difficult to generate a standard mixture of known proteins/peptides to perform such experiments. Both SILAC and label-free approaches have also been suggested to underestimate true fold change and, although to our knowledge no direct comparison between these approaches has been published, it remains that case that iTRAQ underestimation is calculable, linear ([32]; Unwin et al. unpublished data) and, if necessary can be partially corrected. That said, the aim of most iTRAQ experiments is the generation of putative targets for more directed validation experiments and as such the question asked is 'does this protein change?', rather than 'by exactly how much does this protein change? In our hands, however, the vast majority (>95%) of protein changes which we have identified with iTRAQ and gone on to validate with functional assays or Western blot have proven to be correct [35].

It has been reported that iTRAQ labelling can in fact lower the number of peptide identifications generated. Both such studies were performed on an LTQ-Orbitrap mass spectrometer, which employs a slightly different ion path and fragmentation to a standard Quadropole-Time of Flight instrument. In the first study by Pichler et al. [36] identification was achieved by collisionally induced dissociation in the ion trap, which in theory provides more efficient fragmentation but will not provide quantification as products which are less than 1/3 of the precursor ion mass are not retained in the trap (in this case each peptide was also fragmented by Higher energy Collisional Dissociation, or HCD, to generate quantification ions). The lower rate of identifications was partly thought to arise from both the presence of iTRAQ label-generated ions which the analysis software failed to take into account, and partly due to the different physicochemical properties of iTRAQ labelled vs. unlabelled peptides. The second study by Thingholm et al. [37] showed a similar decrease in identification of iTRAQ labelled peptides using this instrument, this time using HCD fragmentation. The authors partly assigned this lower efficiency of analysis to the altered charge state of iTRAQ labelled peptides. They observed (as have we, unpublished data) that while un-labelled peptides generally ionise with 2 or 3 positive charges, iTRAQ labelled peptides carry predominantly 3 or 4 positive charges. Thingholm et al. hypothesised that this could be responsible for the lower rate of identification (since product ion spectra are more complex for 4+ precursor ions than say 2+ ions, since product ions can carry 1, 2 or 3 charges, whereas from a 2+ precursor all products should be 1+). They developed a method of reducing the charge state of the iTRAQ labelled peptides by introducing a basic ammonia vapour into the ion spray. This partly reduced the average charge of the precursor ions and marginally increased the number of peptides successfully identified.

Both studies, however failed to take into account that iTRAQ and non-iTRAQ labelled peptides require different collision energies for optimal fragmentation. In our hands on a AB Sciex QStar Q-ToF instrument, a 4-plex labelled peptide required around 8–12 more eV of energy than its unlabelled equivalent to obtain optimal fragmentation. For 8-plex this number is slightly higher again [38]. We see little or no difference in the rate of peptide identification when samples they are analysed under instrument setup which are optimal for each sample type (unpublished observation).

5.6 Uses of the iTRAQ Technology

Since its introduction in 2004, iTRAQ has been utilised in many settings and for many purposes, including investigation of cell behaviour in model systems, analysis and comparison of primary cell types, temporal regulation of protein expression and modification, and biomarker discovery from primary clinical samples.

Among the first uses of the iTRAQ technology was the analysis of clinical material to look for biomarkers [39]. Since this time, biomarker analysis has been a key use of iTRAQ, with its multiplex nature making it particularly attractive for studies where many samples are analysed. In a study on head and neck squamous cell epithelial carcinomas, Ralhan et al. [40] analysed 15 individual samples and identified and quantified 811 proteins of which 38 differed. They further validated three proteins as plausible cancer biomarkers by western blot, RT-PCR and immunohistochemistry [40, 41]. Chen et al. [42] have identified 55 proteins which were consistently changing in the urine of bladder cancer patients and Rajcevic and colleagues [43] have identified 1,460 proteins with quantification from a rat xenograft model of glioblastomas, a rare cancer with a poor prognosis with patient survival between 12 and 14 months. They went on to use tissue microarrays to validate four of the candidate markers as potentially useful and novel markers for high grade disease.

Clinical biomarker analysis is not limited to cancers. By analysing the amniotic fluid of women who went into premature labour Romero et al. [44] were able to identify 67 proteins which were differentially expressed compared to a control group. Vallelian and colleagues applied iTRAQ proteomics to identify the haemoglobin scavenger receptor CD163 as a target of glucocorticoids in the treatment of autoimmune haemolytic anaemia [45].

This isobaric tagging method has also been successfully applied to studies of basic cell and stem cell biology. Williamson et al. [46] identified and relatively quantified 2,389 proteins in a time course of murine embryonic stem cells differentiation towards the haemangioblast using iTRAQ. Of these 2,389 proteins 1,057 were mapped onto a microarray analysis of the same cells. Poor overall correlation between protein expression changes and changes in levels of mRNA suggested that there was a strong post-translational regulation of protein levels in differentiating stem cells.

Further studies carried out on early differentiating murine embryonic stem cells by Lu et al. [35] were used to evaluate the effect of the downregulation of Nanog, a transcription factor critical for the maintenance of the stem cell phenotype. This paper compared four levels of gene regulation (chromatin configuration change at specific genes, initiation of RNA synthesis, mRNA changes and protein changes) for 1,632 proteins over a 5 day timecourse of Nanog depletion by shRNA. Combining these data demonstrate distinct levels of regulation for key gene families and shows a rapid downregulation of the Nanog protein network, in the absence of changes in gene transcription.

The ability to analyse peptides over a time course, for example, in response to drugs or cellular stimuli in a single experiment is a strength iTRAQ technology has over many other quantitation techniques. It is also particularly appropriate for analysis of complex regulatory events such a post translational modification, the most commonly studied of which is phosphorylation. Zhang et al. identified and quantified 108 tyrosine phosphorlyation sites on 76 proteins which changed in relation to epidermal growth factor receptor stimulation by labelling with an iTRAQ tag there was a timeline established for the phosphorylation profile for each site [47].

The BCR/ABL tyrosine kinase is the product of a genetic translocation and results in chronic myeloid leukaemia (CML). Lee et al. analysed BCR/ABL transduced Ba/F3 cell line and discovered several phosphoproteins which changed on introduction of the BCR/ABL construct using iTRAQ technology [48]. iTRAQ has also been used for the profiling of kinase inhibitor activity in the CML cell line K562, enabling Bantscheff et al. to map phosphorlyation events downstream of several tyrosine kinase inhibitors targets [49].

Other post translational modifications such as glycosylation [50] and nitration [51] have also been studied using iTRAQ technology. Rosenzweig et al. outlined a methodology which could enable researchers to discover several post translational modifications in a high throughput manner [52]. Phosphorlyation of serine and threonine residues were identified and quantified alongside aceylation, methylation, glycoslyation, fucosylation and hexsylation in *Leishmania donovani* intracellualar parasitic development.

Although the majority of iTRAQ studies being undertaken are based on an expression proteomics question there are other interesting applications which have been developed. The determination of a protein's localisation within the cell at a specific point in time can suggest to researchers a particular role that protein may play. Localisation of organelle proteins by isotope tagging (LOPIT) was first conceived by Dunkley et al. [53] in 2004 and has been further developed since [54]. Density gradient centrifugation is performed on samples and organelles will localise at different points along the gradient. The fractions are then iTRAQ labelled according to known organelle localisation, pooled and analysed. The abundance of a protein along the density gradient can be elucidated and it was demonstrated that proteins which are known to exist in specific organelles clustered together enabling the localisation of proteins of unknown location to be determined.

The analysis of protein complexes and their phosphorylation state has also been addressed using iTRAQ technology. Pflieger et al. [55] published a method which enabled them to identify specific interactors of an insulin receptor substrate homolog, Chico. Not only were 14-3-3 proteins identified as interactors of Chico but the phosphorylation status of proteins in the complex could also be determined.

5.7 Summary

In summary, iTRAQ can be employed for a wide range of applications in the field of proteomics. The strength of iTRAQ lies in its versatility; from the analysis of post translational modification to its application to clinical proteomics. As a chemical tag it can be applied to almost any sample, a drawback for some quantification techniques. It is an especially useful method in the analysis of non dividing cells such as stem cells or clinical samples where incorporating iTRAQ tags on up to eight samples allows them to be analysed simultaneously. The technique has been shown numerous times to successfully identify changes in peptide and protein levels in many biological matrices. Concerns have been published regarding the accuracy of this quantification but these issues are surmountable giving a robust platform for biomedicine and molecular cell biology research.

Acknowledgments Work in the authors' laboratory is funded by Leukaemia Lymphoma Research and Cancer Research UK.

References

1. International Human Genome Sequencing Consortium (2004) Finishing the euchromatic sequence of the human genome. Nature 431:931–945
2. Zhu J, He F, Song S, Wang J, Yu J (2008) How many human genes can be defined as housekeeping with current expression data? BMC Genomics 9:172
3. Jensen ON (2004) Modification-specific proteomics: characterization of post-translational modifications by mass spectrometry. Curr Opin Chem Biol 8:33–41
4. Siomi H, Siomi MC (2009) On the road to reading the RNA-interference code. Nature 457:396–404
5. Farley AR, Link AJ (2009) Identification and quantification of protein posttranslational modifications. Methods Enzymol 463:725–763
6. Bantscheff M, Schirle M, Sweetman G, Rick J, Kuster B (2007) Quantitative mass spectrometry in proteomics: a critical review. Anal Bioanal Chem 389:1017–1031
7. Unwin RD, Evans CA, Whetton AD (2006) Relative quantification in proteomics: new approaches for biochemistry. Trends Biochem Sci 31:473–484
8. Thompson A, Schafer J, Kuhn K, Kienle S, Schwarz J, Schmidt G, Neumann T, Johnstone R, Mohammed AK, Hamon C (2003) Tandem mass tags: a novel quantification strategy for comparative analysis of complex protein mixtures by MS/MS. Anal Chem 75:1895–1904
9. Ross PL, Huang YN, Marchese JN, Williamson B, Parker K, Hattan S, Khainovski N, Pillai S, Dey S, Daniels S et al (2004) Multiplexed protein quantitation in Saccharomyces cerevisiae using amine-reactive isobaric tagging reagents. Mol Cell Proteomics 3:1154–1169
10. Choe L, D'Ascenzo M, Relkin NR, Pappin D, Ross P, Williamson B, Guertin S, Pribil P, Lee KH (2007) 8-plex quantitation of changes in cerebrospinal fluid protein expression in subjects undergoing intravenous immunoglobulin treatment for Alzheimer's disease. Proteomics 7:3651–3660
11. Bantscheff M, Boesche M, Eberhard D, Matthieson T, Sweetman G, Kuster B (2008) Robust and sensitive iTRAQ quantification on an LTQ Orbitrap mass spectrometer. Mol Cell Proteomics 7:1702–1713
12. Boja ES, Phillips D, French SA, Harris RA, Balaban RS (2009) Quantitative mitochondrial phosphoproteomics using iTRAQ on an LTQ-Orbitrap with high energy collision dissociation. J Proteome Res 8:4665–4675
13. Zhang Y, Ficarro SB, Li S, Marto JA (2009) Optimized Orbitrap HCD for quantitative analysis of phosphopeptides. J Am Soc Mass Spectrom 20:1425–1434
14. Han H, Pappin DJ, Ross PL, McLuckey SA (2008) Electron transfer dissociation of iTRAQ labeled peptide ions. J Proteome Res 7:3643–3648
15. Phanstiel D, Unwin R, McAlister GC, Coon JJ (2009) Peptide quantification using 8-plex isobaric tags and electron transfer dissociation tandem mass spectrometry. Anal Chem 81:1693–1698
16. Phanstiel D, Zhang Y, Marto JA, Coon JJ (2008) Peptide and protein quantification using iTRAQ with electron transfer dissociation. J Am Soc Mass Spectrom 19:1255–1262
17. Formolo CA, Mintz M, Takanohashi A, Brown KJ, Vanderver A, Halligan B, Hathout Y (2011) Time series proteome profiling. Methods Mol Biol 694:365–377
18. Kruger M, Kratchmarova I, Blagoev B, Tseng YH, Kahn CR, Mann M (2008) Dissection of the insulin signaling pathway via quantitative phosphoproteomics. Proc Natl Acad Sci USA 105:2451–2456
19. Pierce A, Unwin RD, Evans CA, Griffiths S, Carney L, Zhang L, Jaworska E, Lee C-F, Blinco D, Okoniewski MJ et al (2008) Eight-channel iTRAQ enables comparison of the activity of six leukemogenic tyrosine kinases. Mol Cell Proteomics 7:853–863
20. Unwin RD, Smith DL, Blinco D, Wilson CL, Miller CJ, Evans CA, Jaworska E, Baldwin SA, Barnes K, Pierce A et al (2006) Quantitative proteomics reveals posttranslational control as a regulatory factor in primary hematopoietic stem cells. Blood 107:4687–4694

21. Zhang Y, Askenazi M, Jiang J, Luckey CJ, Griffin JD, Marto JA (2010) A robust error model for iTRAQ quantification reveals divergent signaling between oncogenic FLT3 mutants in acute myeloid leukemia. Mol Cell Proteomics 9:780–790
22. McHugh L, Arthur JW (2008) Computational methods for protein identification from mass spectrometry data. PLoS Comput Biol 4:e12
23. Shilov I, Seymour S, Patel A, Loboda A, Tang W, Keating S, Hunter C, Nuwaysir L, Schaeffer D (2007) The Paragon Algorithm, a next generation search engine that uses sequence temperature values and feature probabilities to identify peptides from tandem mass spectra. Mol Cell Proteomics 6:1638–1655
24. Shadforth IP, Dunkley TP, Lilley KS, Bessant C (2005) i-Tracker: for quantitative proteomics using iTRAQ. BMC Genomics 6:145
25. Lin WT, Hung WN, Yian YH, Wu KP, Han CL, Chen YR, Chen YJ, Sung TY, Hsu WL (2006) Multi-Q: a fully automated tool for multiplexed protein quantitation. J Proteome Res 5:2328–2338
26. D'Ascenzo M, Choe L, Lee KH (2008) iTRAQPak: an R based analysis and visualization package for 8-plex isobaric protein expression data. Brief Funct Genomic Proteomic 7:127–135
27. Schwacke JH, Hill EG, Krug EL, Comte-Walters S, Schey KL (2009) iQuantitator: a tool for protein expression inference using iTRAQ. BMC Bioinformatics 10:342
28. Rodríguez-Suárez E, Gubb E, Alzueta IF, Falcón-Pérez JM, Amorim A, Elortza F, Matthiesen R (2010) Virtual Expert Mass Spectrometrist: iTRAQ tool for database-dependent search, quantitation and result storage. Proteomics 10:1545–1556
29. Laderas T, Bystrom C, McMillen D, Fan G, McWeeney S (2007) TandTRAQ: an open-source tool for integrated protein identification and quantitation. Bioinformatics 23:3394–3396
30. Lacerda CM, Xin L, Rogers I, Reardon KF (2008) Analysis of iTRAQ data using Mascot and Peaks quantification algorithms. Brief Funct Genomic Proteomic 7:119–126
31. Park SK, Yates JR 3rd (2010) Census for proteome quantification. Curr Protoc Bioinform Chapter 13:Unit 13.12.1–13.12.11
32. Karp NA, Huber W, Sadowski PG, Charles PD, Hester SV, Lilley KS (2010) Addressing accuracy and precision issues in iTRAQ quantitation. Mol Cell Proteomics 9:1885–1897
33. Ow SY, Salim M, Noirel J, Evans C, Rehman I, Wright PC (2009) ITRAQ underestimation in simple and complex mixtures: "The good, the bad and the ugly". J Proteome Res 8:5347–5355
34. Madler S, Bich C, Touboul D, Zenobi R (2009) Chemical cross-linking with NHS esters: a systematic study on amino acid reactivities. J Mass Spectrom 44:694–706
35. Lu R, Markowetz F, Unwin RD, Leek JT, Airoldi EM, MacArthur BD, Lachmann A, Rozov R, Ma'ayan A, Boyer LA et al (2009) Systems-level dynamic analyses of fate change in murine embryonic stem cells. Nature 462:358–362
36. Pichler P, Kocher T, Holzmann J, Mazanek M, Taus T, Ammerer G, Mechtler K (2010) Peptide labeling with isobaric tags yields higher identification rates using iTRAQ 4-plex compared to TMT 6-plex and iTRAQ 8-plex on LTQ Orbitrap. Anal Chem 82:6549–6558
37. Thingholm TE, Palmisano G, Kjeldsen F, Larsen MR (2010) Undesirable charge-enhancement of isobaric tagged phosphopeptides leads to reduced identification efficiency. J Proteome Res 9:4045–4052
38. Unwin RD, Griffiths JR, Whetton AD (2010) Simultaneous analysis of relative protein expression levels across multiple samples using iTRAQ isobaric tags with 2D nano LC-MS/MS. Nat Protoc 5:1574–1582
39. DeSouza L, Diehl G, Rodrigues MJ, Guo J, Romaschin AD, Colgan TJ, Siu KW (2005) Search for cancer markers from endometrial tissues using differentially labeled tags iTRAQ and cICAT with multidimensional liquid chromatography and tandem mass spectrometry. J Proteome Res 4:377–386
40. Ralhan R, Desouza LV, Matta A, Chandra Tripathi S, Ghanny S, Datta Gupta S, Bahadur S, Siu KW (2008) Discovery and verification of head-and-neck cancer biomarkers by differential

protein expression analysis using iTRAQ labeling, multidimensional liquid chromatography, and tandem mass spectrometry. Mol Cell Proteomics 7:1162–1173

41. Matta A, DeSouza LV, Shukla NK, Gupta SD, Ralhan R, Siu KW (2008) Prognostic significance of head-and-neck cancer biomarkers previously discovered and identified using iTRAQ-labeling and multidimensional liquid chromatography-tandem mass spectrometry. J Proteome Res 7:2078–2087
42. Chen Y-T, Chen C-L, Chen H-W, Chung T, Wu C-C, Chen C-D, Hsu C-W, Chen M-C, Tsui K-H, Chang P-L et al (2010) Discovery of novel bladder cancer biomarkers by comparative urine proteomics using iTRAQ technology. J Proteome Res 9:5803–5815
43. Rajcevic U, Petersen K, Knol JC, Loos M, Bougnaud S, Klychnikov O, Li KW, Pham TV, Wang J, Miletic H et al (2009) ITRAQ-based proteomics profiling reveals increased metabolic activity and cellular cross-talk in angiogenic compared with invasive glioblastoma phenotype. Mol Cell Proteomics 8:2595–2612
44. Romero R, Kusanovic JP, Gotsch F, Erez O, Vaisbuch E, Mazaki-Tovi S, Moser A, Tam S, Leszyk J, Master SR et al (2010) Isobaric labeling and tandem mass spectrometry: a novel approach for profiling and quantifying proteins differentially expressed in amniotic fluid in preterm labor with and without intra-amniotic infection/inflammation. J Matern Fetal Neonatal Med 23:261–280
45. Vallelian F, Schaer CA, Kaempfer T, Gehrig P, Duerst E, Schoedon G, Schaer DJ (2010) Glucocorticoid treatment skews human monocyte differentiation into a hemoglobin-clearance phenotype with enhanced heme-iron recycling and antioxidant capacity. Blood 116:5347–5356.
46. Williamson AJK, Smith DL, Blinco D, Unwin RD, Pearson S, Wilson C, Miller C, Lancashire L, Lacaud G, Kouskoff V et al (2008) Quantitative proteomics analysis demonstrates post-transcriptional regulation of embryonic stem cell differentiation to hematopoiesis. Mol Cell Proteomics 7:459–472
47. Zhang Y, Wolf-Yadlin A, Ross PL, Pappin DJ, Rush J, Lauffenburger DA, White FM (2005) Time-resolved mass spectrometry of tyrosine phosphorylation sites in the epidermal growth factor receptor signaling network reveals dynamic modules. Mol Cell Proteomics 4:1240–1250
48. Lee CF, Griffiths S, Rodríguez-Suárez E, Pierce A, Unwin RD, Jaworska E, Evans CA, Gaskell SJ, Whetton AD (2010) Assessment of downstream effectors of BCR/ABL protein tyrosine kinase using combined proteomic approaches. Proteomics 10:3321–3342
49. Bantscheff M, Eberhard D, Abraham Y, Bastuck S, Boesche M, Hobson S, Mathieson T, Perrin J, Raida M, Rau C et al (2007) Quantitative chemical proteomics reveals mechanisms of action of clinical ABL kinase inhibitors. Nat Biotechnol 25:1035–1044
50. Ueda K, Takami S, Saichi N, Daigo Y, Ishikawa N, Kohno N, Katsumata M, Yamane A, Ota M, Sato T-A et al (2010) Development of serum glycoproteomic profiling technique; simultaneous identification of glycosylation sites and site-specific quantification of glycan structure changes. Mol Cell Proteomics 9:1819–1828
51. Chiappetta G, Corbo C, Palmese A, Marino G, Amoresano A (2009) Quantitative identification of protein nitration sites. Proteomics 9:1524–1537
52. Rosenzweig D, Smith D, Myler PJ, Olafson RW, Zilberstein D (2008) Post-translational modification of cellular proteins during Leishmania donovani differentiation. Proteomics 8:1843–1850
53. Dunkley TPJ, Watson R, Griffin JL, Dupree P, Lilley KS (2004) Localization of organelle proteins by isotope tagging (LOPIT). Mol Cell Proteomics 3:1128–1134
54. Trotter MWB, Sadowski PG, Dunkley TPJ, Groen AJ, Lilley KS (2010) Improved sub-cellular resolution via simultaneous analysis of organelle proteomics data across varied experimental conditions. Proteomics 10:4213–4219
55. Pflieger D, Jünger MA, Müller M, Rinner O, Lee H, Gehrig PM, Gstaiger M, Aebersold R (2008) Quantitative proteomic analysis of protein complexes. Mol Cell Proteomics 7:326–346

Chapter 6
Deciphering Protein Complexes and Protein Interaction Networks for Stem Cell Pluripotency

Jianlong Wang

Abstract Embryonic stem cells (ESCs) hold great promise in regenerative medicine owing to their unique property of unlimited self-renewal while retaining multilineage differentiation capacities. Stem cell biology has been advanced by high throughput genomics and proteomics approaches toward identifying a fuller repertoire of genetic and epigenetic regulatory factors and understanding how they function individually and/or combinatorially in regulating self-renewal and maintaining pluripotency. Proteins function as members of protein complexes and form a myriad of protein-protein interactions in governing proper transcriptional output and cellular identity. Construction of protein-protein interaction networks together with other large datasets such as expression profiles and target gene occupancy is essential in facilitating a comprehensive understanding of the mechanisms of ESC self-renewal and pluripotency. This chapter will summarize current efforts and ongoing progresses in dissecting the protein complexes and mapping the protein interaction networks associated with the major pluripotency factors Nanog, Oct4 and Sox2, and provide guidance for refining the current methodologies and developing new tools for high throughput data generation to further our understanding of stem cell pluripotency.

Keywords ESCs • Self-renewal • Affinity purification • Protein-protein interaction network • Interactome

J. Wang (✉)
Department of Developmental and Regenerative Biology, Black Family Stem Cell Institute, Mount Sinai School of Medicine, Atran Building, AB7-10D 1428 Madison Ave, New York, NY 10029, USA
e-mail: jianlong.wang@mssm.edu

A. Ma'ayan and Ben D. MacArthur (eds.), *New Frontiers of Network Analysis in Systems Biology*, DOI 10.1007/978-94-007-4330-4_6,

6.1 Introduction

Embryonic stem cells (ESCs), derived from pre-implantation blastocyst stage embryos, are endowed with unlimited self-renewal capacity while maintaining multilineage differentiation potential, a property often referred to as pluripotency. The self-renewal and pluripotency characteristics of ESCs make these cells uniquely attractive for cell-based therapies as they will provide unlimited material supply for nearly all types of cells through differentiation, and thus offering great hope in regenerative medicine.

Several transcription factors, notably the homeobox proteins Oct4 [62] and Nanog [13, 58], as well as the HMG box containing Sox2 [3], play fundamental roles in early development and stem cell pluripotency. These key factors act in combination to sustain pluripotency by activating ESC critical factors (including themselves) and repressing differentiation-promoting genes. ESCs are sensitive to the dosage of Nanog [34], Oct4 [64] and Sox2 [8, 43]: enforced expression of Nanog relieves ESCs from the LIF requirement for stem cell maintenance [13], promotes transfer of pluripotency after cell fusion [76], and ensures direct reprogramming of somatic cells to the so-called induced pluripotent stem cells (iPSCs) [77]. In contrast, overexpression of Oct4 drives primitive endoderm differentiation [64], possibly due to direct repression of the Nanog promoter by excessive Oct4 [66]. In addition, a small increase of Sox2 triggers the differentiation of mouse ESCs toward mesoderm and ectoderm lineages [43], presumably through perturbation of Oct4 expression [57]. Dosage sensitivity suggests that the ESC state reflects a balance of multiple transcriptional inputs that are likely exerted through association and dissociation of multiprotein complexes. In specifying lineages, Oct4 and Cdx2 counteract each other's functions to shift the balance between trophectoderm and inner cell mass (ICM) fates [65], and Nanog and Gata6 antagonize each other to define epiblast and primitive endoderm lineages [15]. The ESC state, therefore, is likely to be maintained by the continuous and direct interplay of multiple nuclear factors, acting in cooperative and antagonistic modes.

Recent efforts employing high throughput and genomewide approaches such as microarray [37, 73], chromatin immunoprecipitation [9, 16, 41, 50] and RNAi studies [18, 21, 26, 35] have led to identification of an array of self-renewal regulators and pluripotency factors in ESCs. These studies have enhanced our understanding of how stem cells maintain the unique state of pluripotency and how stem cell-like characteristics can be imposed on somatic cells *via* fusion-based or factor-based reprogramming processes (see review [38]). In a post-genomic era, however, it becomes obvious that the pluripotency machinery of a stem cell is far more complicated than simply the collection of specific transcripts, proteins and target loci of known pluripotency factors. Through protein complex formation, translational and post-translational modifications, and degradation, the functional output of these systems is difficult to predict based solely upon gene/protein expression and/or genomic occupancy. It is clear that consideration of the transcriptome alone offers an incomplete and biased interpretation of the underlying cell biology [33, 52].

In addition, it has been demonstrated that many target loci bound by certain factors in chromatin immunoprecipitations may not have direct biological significance [6, 78, 103]. Vital cellular functions require the coordinated action of a large number of proteins that assemble into an array of multi-protein complexes of distinct composition and structure to regulate the transcription of target genes [16, 41]. A key towards understanding the molecular basis of self-renewal and pluripotency of ESCs lies in mapping the intricate protein interaction networks encompassing many pluripotency factors. Current efforts in identifying the protein constituents of stem cell protein complexes and characterizing the pluripotency interaction networks (interactomes) surrounding several critical pluripotency factors have uncovered new factors in self-renewal signaling pathways and provided a wealth of valuable information on stem cell pluripotency [7, 67, 88, 91]. For this book chapter, I will review the current status and ongoing efforts in deciphering the pluripotency protein complexes and construction of protein interaction networks in ESCs.

6.2 Overview of the Approaches for Affinity Purification of Protein Complexes in ESCs

Mammalian protein complexes have been studied by combining protein affinity purification (AP) with mass spectrometry (MS) and bioinformatics. AP makes use of specific binding interactions between molecules and generally involves three basic steps: (1) incubation and binding reaction; (2) washing to remove nonspecific binding; (3) dissociation and recovery of the bound material. Various AP methods have been developed with modifications to one or few of these three steps to optimize purification of the protein of interest [4, 19].

AP strategies can be broadly classified into two main approaches according to the nature of the target molecule (Fig. 6.1): if the target molecule is the native endogenous protein, then the antibody-based affinity purification is applied; if however, the target molecule is tagged with an epitope such as FLAG [23], which can be introduced into cells either by a knockin strategy or as an ectopic overexpression vector, then affinity purification will be based on the affinity tag. Two (or more) different affinity tags [47] can be used for tandem (sequential) affinity purification, which can increase specificity and reduce the background [72, 97].

There are several advantages associated with the AP-MS method: first, AP-MS can be performed under physiological conditions, in the native organism or cell type; second, it does not typically perturb relevant post-translational modifications (PTMs), which are often crucial for the organization and/or activity of complexes and can also be identified by MS; third, it can be used to probe dynamic changes in the composition of protein complexes when used in combination with quantitative proteomics techniques such as *i*sobaric *t*ag for *r*elative and *a*bsolute *q*uantitation (iTRAQ) [104] and *s*table *i*sotope *l*abeling with *a*mino acids in *c*ell culture (SILAC)[31] (see more in Sect. 6.4.3 and Chap. 5 in this volume).

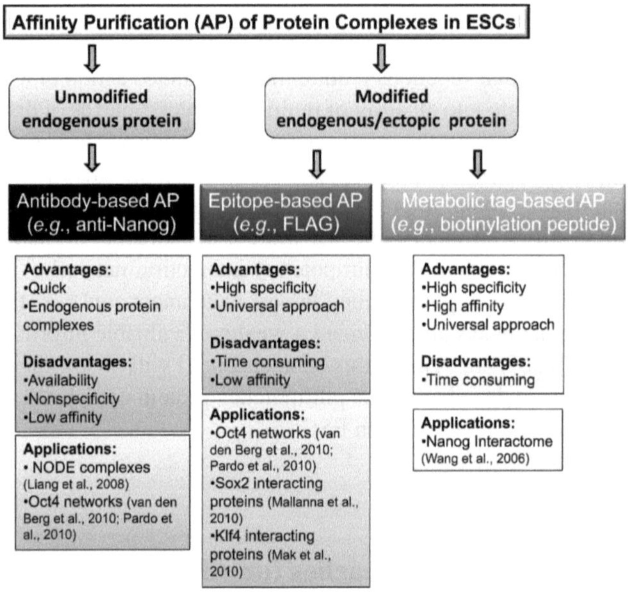

Fig. 6.1 Summary of the strategies for affinity purification of protein complexes in ESCs

6.2.1 Antibody-Based Affinity Purification of Endogenous Protein Complexes

Antibody pull down, also known as immunoprecipitation, is a technique used to isolate a particular protein (and its associated proteins) from solution by means of precipitation. The protein precipitate is formed by coupling the solution with an antibody that specifically binds to the target protein. By isolating a known protein from a complex, other proteins that closely interact with the desired protein may also be pulled out of the complex. Therefore, antibody pull down reveals potential endogenous protein-protein associations. The proteins in a complex may be further separated by SDS polyacrylamide gel electrophoresis (SDS-PAGE) to allow for easy antibody detection of certain known candidate proteins by Western blotting or direct identification of unknown candidate proteins by MS.

The advantages of antibody-based affinity purification are: (1) no transgenic lines need to be established, so the experiment can be done in a speedy manner; and (2) the endogenous protein complexes can be purified, which signifies the biological and functional relevance of the identified interacting proteins. The disadvantages associated with this approach are: (1) most antibodies suffer from non-specific reactivity such that spurious protein complexes will co-purify with the *bona fide* protein complexes; and (2) the affinity between antibody and the target protein is low, and some interactions may be weak and lost during purification procedure. Such a method has been applied to purify endogenous protein complexes of Nanog and Oct4 in mouse ESCs, which yielded a small number of Nanog [48] and Oct4 associated protein complexes [48, 67, 88] and provided limited and yet still valuable information on Nanog and Oct4 function in stem cell pluripotency (see details in Sect. 6.3.3).

6.2.2 *Epitope Tagging for Affinity Purification of Protein Complexes in ESCs*

Among many epitope tags (see review [94]) that are used for AP studies, the FLAG peptides DYKDDDDK and MDYKDDDDK are the most widely used affinity tags for both immunodetection and AP. The FLAG tag can be multiplied (*e.g.*, 3×FLAG) to increase affinity and specificity and can be placed at either the amino-terminus or carboxyl-terminus and in association with other tags for tandem affinity purification. The FLAG tagged protein and its associated protein complexes can be isolated with anti-FLAG antibody (either in a free form or cross linked with gel matrices such as Protein G-Dynal beads) and eluted with FLAG peptides by competition. The 3×FLAG tagging strategy has been employed by the two recent studies to construct an expanded Oct4 interaction network in mouse ESCs [67, 88] (see Sect. 6.3.2). The main advantage of using the epitope tagging for AP is that it makes AP possible for almost all the proteins of interest, particularly for those that antibodies are not available. This makes high throughput analysis of multiple protein complexes possible. In addition, a relatively higher affinity and specificity of the epitope than that of the endogenous antibody also makes it a favorable choice for AP. However, such advantages of affinity and specificity seem to be dependent on cellular context. Even though the FLAG-based AP has been successfully applied in HeLa [12] and HEK293 [1] cells, its application in ESCs still suffers from high background or non-specific binding, presumably due to the presence of proteins in ESC extracts that are reactive nonspecifically to the FLAG and/or relatively low affinity of the epitope tag and the FLAG antibody. This is manifested by the presence of multiple nonspecific species in Western blotting of ESC lysates or nuclear extracts (our unpublished observation) and presence of Oct4 peptides in control samples of the published Oct4 affinity purification study [88]. Therefore, a refined AP strategy employing FLAG in tandem with a second tag is more often a preferred method. Alternatively, a new metabolic biotin tagging strategy has been developed to complement and improve the AP studies in ESCs [40] (see below).

6.2.3 *Metabolic Biotin Tagging for Affinity Purification of Protein Complexes*

In dissecting the pluripotent state, we have employed *in vivo* biotinylation of critical transcription factors including Nanog and Oct4 in mouse ESCs for affinity purification of protein complexes [89] and ChIP-on-chip [40] for target identification. From these data we have constructed protein-protein [91] and protein-DNA [41]

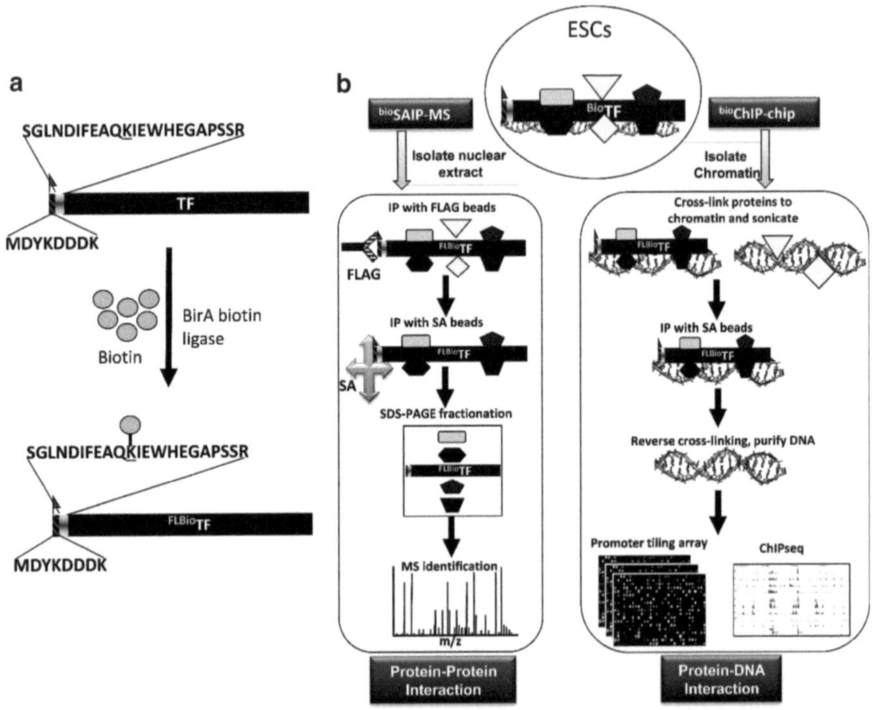

Fig. 6.2 In vivo biotinylation-based strategies for mapping protein-protein interactions in ESCs. (**a**) Modification of a transcription factor (*TF*) with tandem tags (FLAG and Biotinylation Peptide). *E.Coli* BirA biotin ligase catalyzes the addition of biotin to the lysine residue in the biotinylation peptide. (**b**) Affinity methods to capture protein partners (^{bio}SAIP-MS) associated with the biotinylated TF. The ESCs expressing BirA alone will be processed simultaneously as control (not shown). *Triangles* and *diamonds* denote nonspecific binding proteins; other *shapes* denote specific binding proteins

regulatory networks controlling stem cell pluripotency. *In vivo* biotinylation is based on a short 'biotinylation peptide' [75] fused to a protein of interest that serves as an *in vivo* substrate mimic for *E. coli* biotin holoenzyme synthetase (BirA), an enzyme that performs highly selective biotinylation of the fusion protein. In mammalian cells, plasmid expression vectors carrying the biotin-tagged transcription factor (^{bio}TF or ^{FLBio}TF with a Flag-biotin dual tag) and BirA (Fig. 6.2a) can be used to obtain high-level production of soluble ^{bio}TF and BirA proteins, and under appropriate culture conditions, the ^{bio}TF protein produced by this system is completely biotinylated. Studies have documented that biotinylation of a tagged TF does not significantly alter protein interactions, DNA-binding properties *in vivo*, or subnuclear distribution [20]. Therefore, it offers a unique methodology to study protein-protein and protein-DNA interactions simultaneously (Fig. 6.2b).

Biotinylation offers a number of advantages over traditional immunoaffinity approaches for protein complex purification. First, the high affinity of biotin for streptavidin (SA) (10^{-15} M kd) allows efficient purification of the biotinylated protein and associated proteins; second, for ChIP applications, the high affinity allows

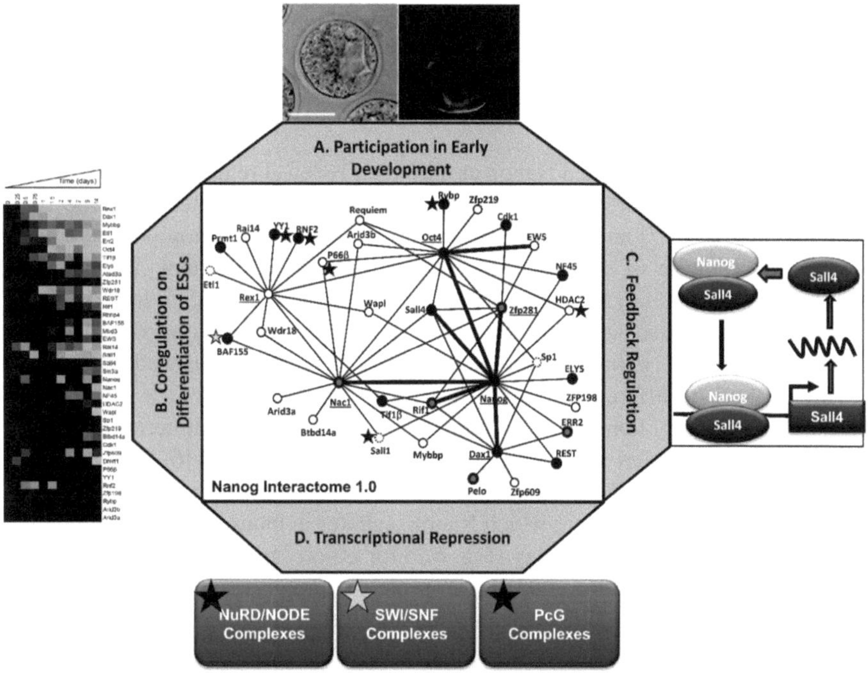

Fig. 6.3 The Nanog interactome for pluripotency of mouse ESCs. Four defining features of the Nanog interactome (labeled **a–d**) for stem cell pluripotency are highlighted. Proteins *underlined* are tagged baits for affinity purification. *Thick lines* indicate interactions confirmed by coIP studies [45, 91, 95]. *Black circles* indicate proteins whose knockout results in defects in proliferation and/or survival of the inner cell mass or other aspects of early development; *gray circles* indicate proteins whose reduction by RNA-mediated interference results in defects in self-renewal and/or differentiation of ESCs; *dotted circles* are proteins whose knockout results in later developmental defects; *white circles* denote proteins for which no loss-of-function data are available

high stringency washing conditions, which reduces background binding that may occur with other affinity tags; third, naturally biotinylated proteins are rare and well-defined [7, 20, 91], and the chance for cross-reaction is quite low; fourth, the approach obviates the need to generate protein-specific antibodies, which often cross-react with other cellular proteins; finally, it is critical that the tagged proteins maintain their functional and structural integrity when expressed. This makes the biotin tag more appealing than other bigger tags (*e.g.*, 3×FLAG) such that structural hindrance from such a small peptide tag is minimal. On the other hand, additional time is required to establish cell lines for *in vivo* biotinylation when compared to direct antibody-based immunoprecipitation, and like epitope tagging, cell lines expressing a controlled level of biotinylated proteins are necessary for analysis [91] since ectopic expression of a protein drastically beyond endogenous levels can result in spurious protein complexes and increases in nonspecific DNA binding. Nevertheless, these concerns are largely outweighed by the superior specificity and highest affinity of the biotin-streptavidin binding which enabled successful construction of "the Nanog interactome" [91] (Fig. 6.3) and an extended transcriptional regulatory network [41] in mouse ESCs.

6.3 Protein Complexes Associated with Nanog, Oct4 and Sox2

6.3.1 The Nanog Interactome

Genetic studies have defined Nanog as a key self renewal regulator that is essential for early development [14, 58] and for ground-state pluripotency of the inner cell mass (ICM) [77] and its *in vitro* derivative ESCs [99]. Nanog is also required for reprogramming of somatic cells to an embryonic pluripotent state [77] and conferring pluripotency to somatic cells upon cell fusion [76]. ESCs lacking Nanog exhibit compromised self-renewal and tend to differentiate toward endodermal lineage. In contrast, enforced expression of Nanog results in enhanced self-renewal at the expense of differentiation propensity [13].

As a divergent homeobox protein, Nanog likely homodimerizes [59, 90] and function in concert with other critical factors such as Oct4 [62] and Sox2 [3]. To further understand the interactive nature of Nanog, we have explored the protein interaction network in which Nanog operates in mouse ESCs. We employed metabolic biotin tagging strategy (Fig. 6.2) for affinity purification of Nanog protein complexes and its associated partner protein complexes followed by MS-based microsequencing [40]. Large-scale purifications were performed with both one-step (streptavidin capture alone) or tandem (FLAG-immunoprecipitation followed by streptavidin capture) [89]. A set of consistent, stringent selection criteria were then applied to each AP-MS experiment to ensure identification of *bona fide* candidate proteins: first, due to the nature of *in vivo* biotinylation, there are background proteins present in both control and tagged samples consisting of mostly naturally biotinylated carboxylases and their associated enzymes as well as some ribosomal proteins. These have been well characterized [20] and thus were removed from the final candidate list; second, some proteins may be identified in both control and tagged samples during the one-step purification. In this case, only candidates with predominantly higher peptide numbers identified by MS in the tagged as compared with the control samples were selected. Alternatively, tandem affinity purification were also performed to confirm such candidates; third, proteins with documented membrane, cytoplasmic, or mitochondrial localization, if present, were excluded; fourth, for proteins specific to tagged samples, only those with ≥ 2 peptides sequenced from at least two independent purifications (either two singles or one single and one tandem) were included in the final candidate list. We have identified a total of 17 proteins of highest confidence that are physically associated with Nanog, either directly or indirectly through other Nanog interacting proteins [91]. In an iterative fashion we then identified partners of several Nanog-associated proteins (including Oct4) and constructed the protein interaction network surrounding Nanog, *i.e.*, the Nanog interactome (Fig. 6.3).

There are four outstanding features associated with the Nanog interactome. First, the network is remarkably enriched for proteins that are required individually to control the survival or differentiation of the ICM or aspects of early embryonic development (Fig. 6.3a). Second, most genes encoding the proteins within the network are co-regulated and, specifically, downregulated during ESC differentiation based on available microarray data from ESC differentiation studies [69] (Fig. 6.3b). Third, when compared with the target genes of Nanog and Oct4 from the ChIP-on-chip [9] or ChIP-PET [50] studies, a notable feedback regulation mode was evident: many (at least 56%) of the genes encoding the proteins of the network [*e.g.*, Sall4 [95]] are putative Nanog and/or Oct4 targets that also serve as "upstream" effectors to control, either positively or negatively, their own transcriptional regulation (Fig. 6.3c). Fourth, a number of factors in the Nanog interactome with both ESC-specific and ubiquitous expression patterns connect to several epigenetic regulatory pathways. These include the histone deacetylase NuRD (P66β and HDAC2), PRC1 (YY1, RNF2/Ring1B and Rybp) and SWI/SNF chromatin remodelling (BAF155) complexes (Fig. 6.3d). The ESC state is marked by open chromatin and hypertranscription [86] such that multiple loci encoding developmental regulators are often associated with bivalent chromatin marks and poised for imminent activation upon differentiation [5]. Therefore, the repressive machinery embedded within the pluripotency Nanog interactome provides a failsafe mechanism to prevent premature expression of key developmental genes under such dynamically open chromatin conformation of ESCs. Taken together, the Nanog interactome illustrates the requirement for both genetic and epigenetic regulatory control of ESC pluripotency [63] and highlights the importance of transcriptional repression for stem cell pluripotency [17].

Functional studies using RNAi by other groups subsequently confirm several proteins within the Nanog interactome for their function in ESC maintenance. For example, Err2 (Esrrb) [36, 50], Rif1 [50], Sall4 [24, 49, 98, 102], Dax1 (Nr0b1) [61, 79] were individually confirmed by other candidate approaches. In addition, genome wide RNAi studies [21, 26, 35] further validate a number of other network proteins for their roles in self-renewal and pluripotency of ESCs. These studies highlight the efficiency and validity of the biotin-mediated AP-MS strategy for studying protein-protein interactions in ESCs. On the other hand, subsequent studies identified additional Nanog interacting proteins that are not present in the Nanog interactome, such as Smad1 [80], NFκB [85] and Med12 [87], suggesting that our AP method is non-saturating. Alternatively, the purification condition, particularly the high salt (350 mM NaCl) used in our initial study may be biased toward purifying stronger interacting proteins. Therefore, future optimization of affinity purification conditions will be needed to maximize AP strategies for proteomic studies in ESCs, and future studies using improved AP conditions or complementary AP approaches may uncover new partners of Nanog and reveal novel functions for Nanog in stem cell pluripotency.

6.3.2 *The Oct4 Interactome*

Oct4 is vital to the development of an embryo during its early stages of differentiation into somatic cells. Early studies found that the loss of Oct4 expression in mouse embryos causes lack of development of the ICM of the zygote [62]. Oct4 interacts with multiple core pluripotency TFs and connects with several epigenetic regulators in the Nanog interactome [91] (Fig. 6.3). Growing evidence suggests that Oct4 is the key player for genetic and epigenetic regulation of stem cell pluripotency. First, Oct4 is an essential factor that functions alone or with other ESC TFs for factor-based somatic cell reprogramming [68, 81, 82, 100]. Second, it cooperates with Sox2 and Nanog to repress *Xist* [60] and interacts with CTCF to activate *Tsix* [22] in coupling X inactivation reprogramming to the control of pluripotency during embryogenesis. Third, it interacts with Eset, the histone H3K9 methyltransferase, to restrict extra-embryonic trophoblast lineage potential in ESCs [101]. Fourth, it controls the chromatin architecture of ESCs through direct regulation of downstream targets encoding histone H3K9 demethylases Jmjd1a and Jmjd2c, which function in part by modulating H3K9 methylation of the pluripotency factors Tcl1 and Nanog, respectively [51].

While the genomic loci occupied by Oct4 have been extensively identified in both mouse [10, 41, 50] and human [9] ESCs, the spectrum of Oct4 interaction partners is underexplored. While the Nanog interactome encompasses Oct4 and several Oct4-interacting proteins, the Oct4-centered protein interaction network had not been fully explored until two recent complementary studies [67, 88]. These two studies both employed the 3×FLAG tag for affinity purification. The van den Berg study [88] took advantage of the ZHBTc4 ESCs [64] and established an ESC line that expresses 3×FLAG-Oct4 in the absence of endogenous Oct4 (Fig. 6.4a). The Pardo study [67] employed BAC transgenesis to introduce a tandem tag (3×FLAG and calmodulin binding peptide-CBP) fused to Oct4 under the endogenous Oct4 regulatory elements (Fig. 6.4) at a sub-endogenous level (30% wt) and integrated it in a precise location in the mouse genome. Again, as we pointed out for the Nanog interactome study, the current two Oct4 network studies are also limited: there are 54 and 92 Oct4-interacting proteins identified, respectively, by the two studies with mere 20 common proteins (Fig. 6.4). This is largely due to the different experimental platforms and conditions employed by the two studies. It is unclear at this point whether the *bona fide* Oct4-centered interactome should be constructed as a union or intersection of the two data sets [46]. Therefore, additional complementary AP-MS studies are needed to resolve the issue. Nevertheless, several consistent features are associated with the two Oct4-centered interactomes: first, the Oct4 interactome is also enriched for factors critically important for ESC maintenance and early development; second, the majority of the genes encoding these two Oct4 interactome proteins are co-regulated, and particularly downregulated upon ESC differentiation; third, expression of the majority of Oct4-interacting proteins is controlled by Oct4 and other key ESC transcription factors, suggestive of the intricate linkage between transcriptional regulatory networks and protein interaction networks; fourth, both Oct4 networks connect with epigenetic regulatory complexes such as NuRD and SWI/SNF (Fig. 6.4b), which is consistent with the findings in the Nanog interactome (Fig. 6.3). Overall, these data suggest that the Nanog and Oct4 interactomes are inherently connected. Although the detection of Nanog in the two Oct4 interactome studies is either not present [88] or minimal [67], Oct4 has been confirmed to be one of the interacting proteins of Nanog in the Nanog interactome [91] and in the endogenous Nanog protein complexes [48].

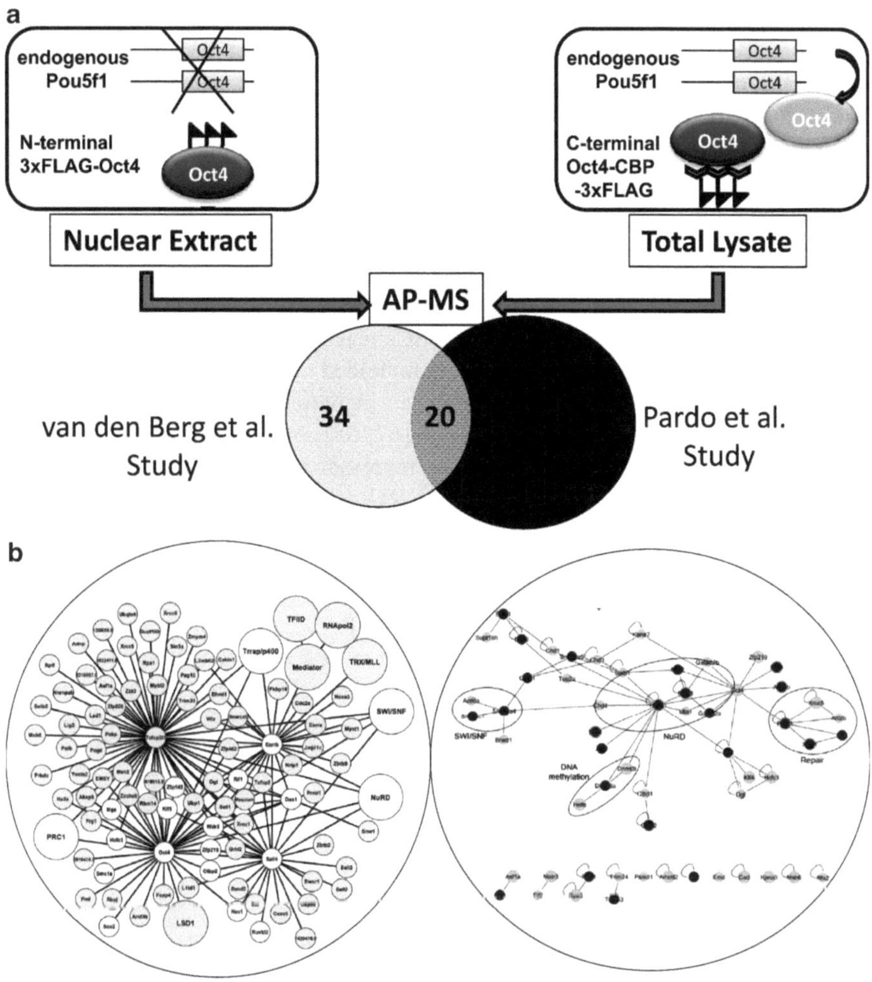

Fig. 6.4 The Oct4 interactome for pluripotency of mouse ESCs. (**a**) Summary of the strategies and results of the two Oct4 network studies [67, 88]. Note the limited overlap between the two studies. (**b**) Network presentation of the two Oct4 interactomes (Reproduced from Cell Stem Cell with permission)

6.3.3 The Endogenous Nanog and Oct4 Protein Complexes

Due to their essential roles in regulating self-renewal and pluripotency of ESCs, the two homeodomain transcription factors Nanog and Oct4 are under intensive study and their endogenous protein complexes have also been sought by antibody-based affinity purification strategies [48, 67, 88]. The study by Liang et al. [48] showed that both Nanog and Oct4 interact with each other and associate with proteins from multiple repression complexes, such as NuRD, Sin3A and Pml complexes.

Surprisingly, of the various core components in the NuRD complex with which Nanog and Oct4 interact, Mta1 was preferred, whereas Mbd3 and Rbbp7 were either absent or present at a sub-stoichiometric levels [48]. The so-called Nanog and Oct4 associated deacetylase (NODE) contains histone deacetylase (HDAC) activity that is comparable to NuRD, and retains its association with Nanog and Oct4 in Mbd3$^{-/-}$ ESCs [48]. However, the NODE complex may simply be the byproduct of the insensitivity of the affinity approach and suboptimal immunoprecipitation condition employed in that study, which was manifested by the limited number of Nanog- and Oct4-interacting proteins identified [48]. The association of Nanog and Oct4 with the conventional NuRD rather than the peculiar NODE repression complex in ESCs was supported by the two independent Oct4 immunoprecipitation studies that yielded a greater number of candidate interacting proteins in the Oct4 complexes including Mbd3 and several other NuRD components with high confidence [67, 88]. Nevertheless, it is clear that Nanog and Oct4 associate with multiple repressor complexes to regulate target gene expression and control ESC fate [48, 91].

The number of candidate interacting proteins identified with such an antibody-based IP approach from all the three studies (Fig. 6.5) is much smaller than that from the epitope tagging strategies (Fig. 6.4), due likely to aforementioned inherent limitation associated with the antibody-based AP, *i.e.*, low affinity and high non-specificity (Fig. 6.1). Therefore, it is important to bear in mind that antibody-based AP should serve only as a complementary approach, but not a gold standard, for identification and validation of potential interacting proteins of a certain protein of interest.

6.3.4 The Sox2 Protein Complexes in ESCs

Although Sox2 is one of the core ESC factors (together with Nanog and Oct4), the mechanism by which Sox2 controls the fate of ESCs is much less well defined. Like Nanog and Oct4, Sox2 expression is also dosage sensitive for stem cell maintenance [43], which highlights the need to decipher Sox2 protein complexes and its interaction network in understanding stem cell pluripotency. Sox2 is well known for its close partnership with Oct4 in target gene regulation [74, 93], however, the evidence for their physical association in protein complexes remains elusive. In one study, Sox2 has been co-purified with Oct4 as part of an expanded Oct4 interaction network [88], however, Sox2 has not been identified in another related Oct4 network study [67] or in the Nanog interactome containing Oct4 [91]. This may reflect the non-saturating nature of affinity purification approaches or the weak interaction between Oct4 and Sox2. Alternatively, there could be only substoichiometric levels of Sox2 in the Oct4 protein complexes and *vice versa*.

Small increase of Sox2 expression level in ESCs promotes their differentiation [43], which compromises the epitope-based AP strategy with ectopic expression of tagged Sox2 in ESCs. Therefore, future endeavor to dissect Sox2 protein complexes in ESCs will require careful manipulation of the ectopic expression of

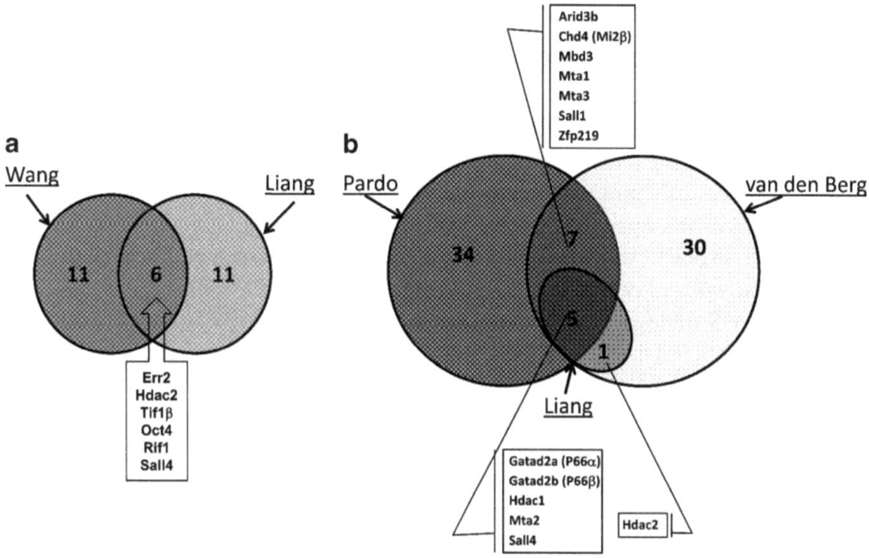

Fig. 6.5 Comparison of the Nanog and Oct4 interacting proteins identified by multiple studies. (**a**) Nanog-interacting proteins; (**b**) Oct4 interacting proteins. Data presented are from four published studies: Wang [91], Liang [48], Pardo [67] and van den Berg [88]

the epitope-tagged Sox2 or adoption of a strategy employing the transgenic rescue of the Sox2 knockout ESCs with the epitope-tagged Sox2. Interestingly, the proteomics studies of Sox2 during early differentiation of ESCs has identified >60 nuclear proteins that associate with Sox2, and a significant number of the identified Sox2-associated proteins also interact with Oct4 and Nanog [56]. While the study may have captured many Sox2-interacting proteins that are required for stem cell maintenance and yet are still active during early stage of differentiation, future studies purifying Sox2 complexes and mapping the Sox2-centered interactome in ESCs are needed to gain a comprehensive understanding of Sox2 function in regulating genes required for self-renewal and pluripotency of ESCs.

6.4 Future Directions

Stem cell research has evolved in the post-genomics era with increasing application of proteomics approaches [92] and high throughput systems biology approaches to define lists of molecular "parts" and regulatory interactions between the "parts" in both undifferentiated ESCs and their differentiated progenies [53, 54, 96]. The analysis of protein complexes and protein-protein interactions is essential for understanding mechanisms of ESC self-renewal and pluripotency. Deploying MS-based proteomics approaches such as AP-MS to decipher protein complexes and dissect protein interaction networks surrounding key pluripotency transcription factors will likely be one of the major scientific inquires in the near future.

6.4.1 High Throughput Experimental System for Network Reconstruction in ESCs

Proteins are much more chemically and structurally diverse than their nucleic acid counterparts, making them intrinsically less suitable for large-scale high-throughput analyses. Large scale protein-protein interaction studies using a high throughput and systematic AP-MS approach were largely confined to cellular systems that are amenable to efficient transfection or genetic engineering such as *Escherichia coli* [11], *Saccharomyces cerevisiae* [27, 44] and human HEK293 cells [25, 29, 39]. The current AP-MS strategies (one gene-one protein-one ESC line) employed in the ESC research [67, 88, 91] are still quite low throughput and time consuming. Considering the exponential growth of the "part-lists" of stem cell factors identified during the past few years owing to the development of microarrays and deep sequencing as well as genome wide RNAi screens, improved AP-MS strategies will have to be developed to meet the demand of high throughput network reconstruction. Future efforts should be directed to develop an integrated strategy for AP-MS in ESCs with the following features: (1) rapid generation of multiple ESC lines bearing epitope tagged baits of interests. Combined lentiviral technology with Gateway cloning technology will expedite cloning of cDNA compatible with publicly and commercially available cDNA libraries and establish stable ESC lines by lentivirus infection [55]; (2) increased yields in protein complex preparation for affinity purification. Due to the dosage sensitivity of many key stem cell factors, the ectopic expression of epitope tagged baits is controlled in a minimal level to preserve the functional integrity of ESCs and avoid formation of the spurious protein complexes due to overexpression. Therefore the total amount of the protein complexes associated with the tagged baits is also limited. A combined vector system allowing tagged bait cDNA expression with simultaneous knockdown of the endogenous protein will be one of the options for increasing recovery of the protein complexes associated with the bait. The added bonus of this transgenic rescue strategy is that the functionality of the tagged bait is also confirmed; (3) direct liquid chromatography tandem MS (LC-MS/MS) analysis of purified complexes to improve the sensitivity and reproducibility of protein identification. The current AP-MS workflow with SDS-PAGE fractionation before MS analysis (Fig. 6.2b) presents a major experimental bottleneck in large-scale and high throughput studies of the protein complexes in ESCs. Development of a strategy to remove endogenously biotinylated proteins (the major background proteins of high abundance in the *in vivo* biotinylation system) and optimization of the affinity purification condition for chemical compatibility of the sample with subsequent LC-MS/MS will be necessary to reach this goal.

6.4.2 Protein-Protein Interactions and Stem Cell Heterogeneity

Heterogeneity is a hallmark of ESCs that might have evolved as a mechanism that enables stem cells to respond to differentiation-inducing signals while retaining their self-renewal potential [30]. ESCs under undifferentiated culture condition show fluctuating expression levels of a number of genes including stem cell specific factors such as Nanog, Rex1, Dppa3 (Stella), Pecam1, Zscan4 and genes normally associated with cell differentiation such as Brachyury/T and Twist2 (see review [83]). The current methodology in mapping the interactome only averages interaction profiles over a large quantity of individual cells with heterogeneous or stochastic expression of some network proteins such as Nanog and Rex1 in the Nanog interactome. Therefore, it is quite possible that, although the extensive protein interactions will likely be present in individual cells, not all of the identified interactions operate within any given individual cell. For example, in pluripotent Nanog null ESCs [14] the Nanog interactome may be compensated by other factors and the ESC state is stabilized by other interactomes such as the Oct4 interactome. It is not currently possible to construct interactomes at the single cell level; however, one could interrogate limited protein-protein interactions from the existing interactome using fluorescence resonance energy transferor related techniques combined with imaging to gain insights into protein-protein interactions and functional stem cell heterogeneity in single cells. In this regard, implementation and improvement of single cell analysis platforms [32, 84] (also see Chap. 1 Review by Arai in this book) will be the necessary step toward this goal.

6.4.3 *Quantitative Protein-Protein Interactions for Stem Cell Pluripotency*

ESCs are sensitive to the dosage of Nanog [34], Oct4 [64] and Sox2 [8, 43]. As a key pluripotency factor, Nanog has been demonstrated to form dimers [59, 90], and Nanog dimerization is critical for interaction with multiple pluripotency network proteins [90]. However, current interactome studies cannot distinguish the mode of action of these dosage sensitive TFs relative to other factors, and it is also unknown whether the monomers or dimers of Nanog form different protein complexes in controlling subset of genes important for the pluripotent state. In addition, Oct4 and Sox2 often act together to regulate a subset of target genes, however, the stoichiometric level of each partner in the Oct4-Sox2 protein complexes cannot be defined by current AP-MS approaches. Reprogramming somatic cells to naïve pluripotency is associated with erasure of epigenetic memory [42, 71] concomitant with gradual increase of Nanog, Oct4 and Sox2 expression and consolidation of the core transcriptional regulatory network [71] that is interdependent on the pluripotency interactome. How dosage sensitivity alters composition of protein complexes and transcriptional regulation of genes controlling stem cell pluripotency remains to be explored. The future efforts to refine the biochemical approaches to isolate distinct protein complexes in combination with quantitative MS technology such as iTRAQ for relative quantitation [70, 104] (see Chap. 5 in this volume for more on iTRAQ) and MRM/MS (*m*ultiple *r*eaction *m*onitoring-MS) [2, 28] for absolute quantitation will be necessary to decipher quantitative relationship of the interacting proteins. In particular, the MRM/MS assay quantifies a specific tryptic peptide that is selected as a stoichiometric representative of the cleaved protein against an internal synthetic stable isotope-labeled peptide, allowing for the absolute measurement of protein concentration [2, 28].

6.4.4 Interactome Dynamics During Stem Cell Fate Changes

The pluripotency interactome so far provides static "snapshots" of undifferentiated ESC state. The exciting therapeutic and regenerative potential of ESCs will only be realized during their fate change to generate more differentiated progenies. Molecular regulation of stem cell fate entails a complex and coordinated action among multilayered regulatory pathways (DNA, mRNA/miRNA and protein) that eventually converge in dynamic protein expression and intricate protein interactions governing distinct cellular identity. Therefore, the interactomes during stem cell fate change are dynamic, and methodologies need to be developed in the future to measure interactome dynamics during cell fate change so as to maximize generation of specific cell types for therapeutic application. One recent study has provided a powerful experimental system to interrogate the Nanog interactome dynamics after shRNA-mediated downregulation of Nanog in ESCs [52]. In addition, another study employed an inducible system to drive Sox2 overexpression and monitor the composition of ectopic Sox2-associated protein complexes during early stage of ESC differentiation, which identified another Sox family protein, Sox21, as a novel regulator for stem cell fate [56]. However, a direct comparison with Sox2 network in the pluripotent state and its dynamics during cell fate change is not possible due to the lack of the Sox2 interactome in ESCs.

6.5 Concluding Remarks

Recent efforts in systematically profiling gene expression in ESCs and global genome-wide functional RNAi screens [18, 21, 26, 35] have yielded a wealth of high throughput data and provided a long molecular 'parts list' of regulatory factors important for ESC self-renewal and pluripotency. How these ever-growing 'molecular building blocks' are interconnected into functional regulatory networks that are ultimately responsible for self-renewal and differentiation of ESCs is unclear. The proteomic studies deciphering protein complexes and protein-protein interactions in ESCs will continue to be an area of active research that likely will bear fruit in the near future.

Acknowledgement The author thanks the editors of this book chapter and all members in the lab for critical reading of the manuscript. The author's current research is supported by a Seed Fund from the Black Family Stem Cell Institute at Mount Sinai School of Medicine, an IDEA grant (C026420) from the Empire State Stem Cell Fund through New York State Department of Health (NYSTEM), and a grant from the NIH (1R01-GM095942-01A1).

Notes While this book chapter is in the final production stage, a more sophisticated Oct 4 interactome has been published by the author's group (Ding et al., Cell Research 22:155–167, 2012; PMID 22083510).

References

1. Adams DG, Wadzinski BE (2007) Isolation and characterization of PP2A holoenzymes containing FLAG-tagged B subunits. Methods Mol Biol (Clifton, NJ) 365:101–111
2. Anderson L, Hunter CL (2006) Quantitative mass spectrometric multiple reaction monitoring assays for major plasma proteins. Mol Cell Proteomics 5:573–588
3. Avilion AA, Nicolis SK, Pevny LH, Perez L, Vivian N, Lovell-Badge R (2003) Multipotent cell lineages in early mouse development depend on SOX2 function. Genes Dev 17:126–140
4. Bauer A, Kuster B (2003) Affinity purification-mass spectrometry. Powerful tools for the characterization of protein complexes. Eur J Biochem 270:570–578
5. Bernstein BE, Mikkelsen TS, Xie X, Kamal M, Huebert DJ, Cuff J, Fry B, Meissner A, Wernig M, Plath K et al (2006) A bivalent chromatin structure marks key developmental genes in embryonic stem cells. Cell 125:315–326
6. Birney E, Stamatoyannopoulos JA, Dutta A, Guigo R, Gingeras TR, Margulies EH, Weng Z, Snyder M, Dermitzakis ET, Thurman RE et al (2007) Identification and analysis of functional elements in 1% of the human genome by the ENCODE pilot project. Nature 447:799–816
7. Blelloch R, Wang Z, Meissner A, Pollard S, Smith A, Jaenisch R (2006) Reprogramming efficiency following somatic cell nuclear transfer is influenced by the differentiation and methylation state of the donor nucleus. Stem Cells (Dayton, Ohio) 24:2007–2013
8. Boer B, Kopp J, Mallanna S, Desler M, Chakravarthy H, Wilder PJ, Bernadt C, Rizzino A (2007) Elevating the levels of Sox2 in embryonal carcinoma cells and embryonic stem cells inhibits the expression of Sox2:Oct-3/4 target genes. Nucleic Acids Res 35:1773–1786
9. Boyer LA, Lee TI, Cole MF, Johnstone SE, Levine SS, Zucker JP, Guenther MG, Kumar RM, Murray HL, Jenner RG et al (2005) Core transcriptional regulatory circuitry in human embryonic stem cells. Cell 122:947–956
10. Boyer LA, Plath K, Zeitlinger J, Brambrink T, Medeiros LA, Lee TI, Levine SS, Wernig M, Tajonar A, Ray MK et al (2006) Polycomb complexes repress developmental regulators in murine embryonic stem cells. Nature 441:349–353
11. Butland G, Peregrin-Alvarez JM, Li J, Yang W, Yang X, Canadien V, Starostine A, Richards D, Beattie B, Krogan N et al (2005) Interaction network containing conserved and essential protein complexes in Escherichia coli. Nature 433:531–537
12. Carey MF, Peterson CL, Smale ST (2010) Purification of mediator from HeLa cell lines expressing a flag-tagged mediator subunit. Cold Spring Harb Protoc 2010:pdb prot5451
13. Chambers I, Colby D, Robertson M, Nichols J, Lee S, Tweedie S, Smith A (2003) Functional expression cloning of Nanog, a pluripotency sustaining factor in embryonic stem cells. Cell 113:643–655
14. Chambers I, Silva J, Colby D, Nichols J, Nijmeijer B, Robertson M, Vrana J, Jones K, Grotewold L, Smith A (2007) Nanog safeguards pluripotency and mediates germline development. Nature 450:1230–1234
15. Chazaud C, Yamanaka Y, Pawson T, Rossant J (2006) Early lineage segregation between epiblast and primitive endoderm in mouse blastocysts through the Grb2-MAPK pathway. Dev Cell 10:615–624
16. Chen X, Xu H, Yuan P, Fang F, Huss M, Vega VB, Wong E, Orlov YL, Zhang W, Jiang J et al (2008) Integration of external signaling pathways with the core transcriptional network in embryonic stem cells. Cell 133:1106–1117

17. Cheong CY, Lufkin T (2010) Transcriptional repression in ES cells. J Cell Biochem 110:288–293
18. Chia NY, Chan YS, Feng B, Lu X, Orlov YL, Moreau D, Kumar P, Yang L, Jiang J, Lau MS et al (2010) A genome-wide RNAi screen reveals determinants of human embryonic stem cell identity. Nature 468:316–320
19. Collins MO, Choudhary JS (2008) Mapping multiprotein complexes by affinity purification and mass spectrometry. Curr Opin Biotechnol 19:324–330
20. de Boer E, Rodriguez P, Bonte E, Krijgsveld J, Katsantoni E, Heck A, Grosveld F, Strouboulis J (2003) Efficient biotinylation and single-step purification of tagged transcription factors in mammalian cells and transgenic mice. Proc Natl Acad Sci USA 100:7480–7485
21. Ding L, Paszkowski-Rogacz M, Nitzsche A, Slabicki MM, Heninger AK, de Vries I, Kittler R, Junqueira M, Shevchenko A, Schulz H et al (2009) A genome-scale RNAi screen for Oct4 modulators defines a role of the Paf1 complex for embryonic stem cell identity. Cell Stem Cell 4:403–415
22. Donohoe ME, Silva SS, Pinter SF, Xu N, Lee JT (2009) The pluripotency factor Oct4 interacts with Ctcf and also controls X-chromosome pairing and counting. Nature 460:128–132
23. Einhauer A, Jungbauer A (2001) The FLAG peptide, a versatile fusion tag for the purification of recombinant proteins. J Biochem Biophys Methods 49:455–465
24. Elling U, Klasen C, Eisenberger T, Anlag K, Treier M (2006) Murine inner cell mass-derived lineages depend on Sall4 function. Proc Natl Acad Sci USA 103:16319–16324
25. Ewing RM, Chu P, Elisma F, Li H, Taylor P, Climie S, McBroom-Cerajewski L, Robinson MD, O'Connor L, Li M et al (2007) Large-scale mapping of human protein-protein interactions by mass spectrometry. Mol Syst Biol 3:89
26. Fazzio TG, Huff JT, Panning B (2008) An RNAi screen of chromatin proteins identifies Tip60-p400 as a regulator of embryonic stem cell identity. Cell 134:162–174
27. Gavin AC, Bosche M, Krause R, Grandi P, Marzioch M, Bauer A, Schultz J, Rick JM, Michon AM, Cruciat CM et al (2002) Functional organization of the yeast proteome by systematic analysis of protein complexes. Nature 415:141–147
28. Gerber SA, Rush J, Stemman O, Kirschner MW, Gygi SP (2003) Absolute quantification of proteins and phosphoproteins from cell lysates by tandem MS. Proc Natl Acad Sci USA 100:6940–6945
29. Glaser S, Schaft J, Lubitz S, Vintersten K, van der Hoeven F, Tufteland KR, Aasland R, Anastassiadis K, Ang SL, Stewart AF (2006) Multiple epigenetic maintenance factors implicated by the loss of Mll2 in mouse development. Development 133:1423–1432
30. Graf T, Stadtfeld M (2008) Heterogeneity of embryonic and adult stem cells. Cell Stem Cell 3:480–483
31. Gruhler S, Kratchmarova I (2008) Stable isotope labeling by amino acids in cell culture (SILAC). Methods Mol Biol (Clifton, NJ) 424:101–111
32. Guo G, Huss M, Tong GQ, Wang C, Li Sun L, Clarke ND, Robson P (2010) Resolution of cell fate decisions revealed by single-cell gene expression analysis from zygote to blastocyst. Dev Cell 18:675–685
33. Gygi SP, Rochon Y, Franza BR, Aebersold R (1999) Correlation between protein and mRNA abundance in yeast. Mol Cell Biol 19:1720–1730
34. Hatano SY, Tada M, Kimura H, Yamaguchi S, Kono T, Nakano T, Suemori H, Nakatsuji N, Tada T (2005) Pluripotential competence of cells associated with Nanog activity. Mech Dev 122:67–79
35. Hu G, Kim J, Xu Q, Leng Y, Orkin SH, Elledge SJ (2009) A genome-wide RNAi screen identifies a new transcriptional module required for self-renewal. Genes Dev 23:837–848
36. Ivanova N, Dobrin R, Lu R, Kotenko I, Levorse J, DeCoste C, Schafer X, Lun Y, Lemischka IR (2006) Dissecting self-renewal in stem cells with RNA interference. Nature 442:533–538

37. Ivanova NB, Dimos JT, Schaniel C, Hackney JA, Moore KA, Lemischka IR (2002) A stem cell molecular signature. Science (New York, NY) 298:601–604
38. Jaenisch R, Young R (2008) Stem cells, the molecular circuitry of pluripotency and nuclear reprogramming. Cell 132:567–582
39. Jeronimo C, Forget D, Bouchard A, Li Q, Chua G, Poitras C, Therien C, Bergeron D, Bourassa S, Greenblatt J et al (2007) Systematic analysis of the protein interaction network for the human transcription machinery reveals the identity of the 7SK capping enzyme. Mol Cell 27:262–274
40. Kim J, Cantor AB, Orkin SH, Wang J (2009) Use of in vivo biotinylation to study protein-protein and protein-DNA interactions in mouse embryonic stem cells. Nat Protoc 4:506–517
41. Kim J, Chu J, Shen X, Wang J, Orkin SH (2008) An extended transcriptional network for pluripotency of embryonic stem cells. Cell 132:1049–1061
42. Kim K, Doi A, Wen B, Ng K, Zhao R, Cahan P, Kim J, Aryee MJ, Ji H, Ehrlich LI et al (2010) Epigenetic memory in induced pluripotent stem cells. Nature 467:285–290
43. Kopp JL, Ormsbee BD, Desler M, Rizzino A (2008) Small increases in the level of Sox2 trigger the differentiation of mouse embryonic stem cells. Stem Cells (Dayton, Ohio) 26:903–911
44. Krogan NJ, Cagney G, Yu H, Zhong G, Guo X, Ignatchenko A, Li J, Pu S, Datta N, Tikuisis AP et al (2006) Global landscape of protein complexes in the yeast Saccharomyces cerevisiae. Nature 440:637–643
45. Lee J, Rhee BK, Bae GY, Han YM, Kim J (2005) Stimulation of Oct-4 activity by Ewing's sarcoma protein. Stem Cells (Dayton, Ohio) 23:738–751
46. Lemischka IR (2010) Hooking up with Oct4. Cell Stem Cell 6:291–292
47. Li Y (2010) Commonly used tag combinations for tandem affinity purification. Biotechnol Appl Biochem 55:73–83
48. Liang J, Wan M, Zhang Y, Gu P, Xin H, Jung SY, Qin J, Wong J, Cooney AJ, Liu D et al (2008) Nanog and Oct4 associate with unique transcriptional repression complexes in embryonic stem cells. Nat Cell Biol 10:731–739
49. Lim CY, Tam WL, Zhang J, Ang HS, Jia H, Lipovich L, Ng HH, Wei CL, Sung WK, Robson P et al (2008) Sall4 regulates distinct transcription circuitries in different blastocyst-derived stem cell lineages. Cell Stem Cell 3:543–554
50. Loh YH, Wu Q, Chew JL, Vega VB, Zhang W, Chen X, Bourque G, George J, Leong B, Liu J et al (2006) The Oct4 and Nanog transcription network regulates pluripotency in mouse embryonic stem cells. Nat Genet 38:431–440
51. Loh YH, Zhang W, Chen X, George J, Ng HH (2007) Jmjd1a and Jmjd2c histone H3 Lys 9 demethylases regulate self-renewal in embryonic stem cells. Genes Dev 21:2545–2557
52. Lu R, Markowetz F, Unwin RD, Leek JT, Airoldi EM, MacArthur BD, Lachmann A, Rozov R, Ma'ayan A, Boyer LA et al (2009) Systems-level dynamic analyses of fate change in murine embryonic stem cells. Nature 462:358–362
53. MacArthur BD, Ma'ayan A, Lemischka IR (2008) Toward stem cell systems biology: from molecules to networks and landscapes. Cold Spring Harb Symp Quant Biol 73:211–215
54. Macarthur BD, Ma'ayan A, Lemischka IR (2009) Systems biology of stem cell fate and cellular reprogramming. Nat Rev Mol Cell Biol 10:672–681
55. Mak AB, Ni Z, Hewel JA, Chen GI, Zhong G, Karamboulas K, Blakely K, Smiley S, Marcon E, Roudeva D et al (2010) A lentiviral functional proteomics approach identifies chromatin remodeling complexes important for the induction of pluripotency. Mol Cell Proteomics 9:811–823
56. Mallanna SK, Ormsbee BD, Iacovino M, Gilmore JM, Cox JL, Kyba M, Washburn MP, Rizzino A (2010) Proteomic analysis of Sox2-associated proteins during early stages of mouse embryonic stem cell differentiation identifies Sox21 as a novel regulator of stem cell fate. Stem Cells (Dayton, Ohio) 28:1715–1727

57. Masui S, Nakatake Y, Toyooka Y, Shimosato D, Yagi R, Takahashi K, Okochi H, Okuda A, Matoba R, Sharov AA et al (2007) Pluripotency governed by Sox2 via regulation of Oct3/4 expression in mouse embryonic stem cells. Nat Cell Biol 9:625–635
58. Mitsui K, Tokuzawa Y, Itoh H, Segawa K, Murakami M, Takahashi K, Maruyama M, Maeda M, Yamanaka S (2003) The homeoprotein Nanog is required for maintenance of pluripotency in mouse epiblast and ES cells. Cell 113:631–642
59. Mullin NP, Yates A, Rowe AJ, Nijmeijer B, Colby D, Barlow PN, Walkinshaw MD, Chambers I (2008) The pluripotency rheostat Nanog functions as a dimer. Biochem J 411:227–231
60. Navarro P, Chambers I, Karwacki-Neisius V, Chureau C, Morey C, Rougeulle C, Avner P (2008) Molecular coupling of Xist regulation and pluripotency. Science (New York, NY) 321:1693–1695
61. Niakan KK, Davis EC, Clipsham RC, Jiang M, Dehart DB, Sulik KK, McCabe ER (2006) Novel role for the orphan nuclear receptor Dax1 in embryogenesis, different from steroidogenesis. Mol Genet Metab 88:261–271
62. Nichols J, Zevnik B, Anastassiadis K, Niwa H, Klewe-Nebenius D, Chambers I, Scholer H, Smith A (1998) Formation of pluripotent stem cells in the mammalian embryo depends on the POU transcription factor Oct4. Cell 95:379–391
63. Niwa H (2007) How is pluripotency determined and maintained? Development 134:635–646
64. Niwa H, Miyazaki J, Smith AG (2000) Quantitative expression of Oct-3/4 defines differentiation, dedifferentiation or self-renewal of ES cells. Nat Genet 24:372–376
65. Niwa H, Toyooka Y, Shimosato D, Strumpf D, Takahashi K, Yagi R, Rossant J (2005) Interaction between Oct3/4 and Cdx2 determines trophectoderm differentiation. Cell 123:917–929
66. Pan G, Li J, Zhou Y, Zheng H, Pei D (2006) A negative feedback loop of transcription factors that controls stem cell pluripotency and self-renewal. FASEB J 20:1730–1732
67. Pardo M, Lang B, Yu L, Prosser H, Bradley A, Babu MM, Choudhary J (2010) An expanded Oct4 interaction network: implications for stem cell biology, development, and disease. Cell Stem Cell 6:382–395
68. Park IH, Zhao R, West JA, Yabuuchi A, Huo H, Ince TA, Lerou PH, Lensch MW, Daley GQ (2008) Reprogramming of human somatic cells to pluripotency with defined factors. Nature 451:141–146
69. Perez-Iratxeta C, Palidwor G, Porter CJ, Sanche NA, Huska MR, Suomela BP, Muro EM, Krzyzanowski PM, Hughes E, Campbell PA et al (2005) Study of stem cell function using microarray experiments. FEBS Lett 579:1795–1801
70. Pewsey E, Bruce C, Tonge P, Evans C, Ow SY, Georgiou AS, Wright PC, Andrews PW, Fazeli A (2010) Nuclear proteome dynamics in differentiating embryonic carcinoma (NTERA-2) cells. J Proteome Res 9:3412–3426
71. Polo JM, Liu S, Figueroa ME, Kulalert W, Eminli S, Tan KY, Apostolou E, Stadtfeld M, Li Y, Shioda T et al (2010) Cell type of origin influences the molecular and functional properties of mouse induced pluripotent stem cells. Nat Biotechnol 28:848–855
72. Puig O, Caspary F, Rigaut G, Rutz B, Bouveret E, Bragado-Nilsson E, Wilm M, Seraphin B (2001) The tandem affinity purification (TAP) method: a general procedure of protein complex purification. Methods (San Diego, Calif) 24:218–229
73. Ramalho-Santos M, Yoon S, Matsuzaki Y, Mulligan RC, Melton DA (2002) "Stemness": transcriptional profiling of embryonic and adult stem cells. Science (New York, NY) 298:597–600
74. Rodda DJ, Chew JL, Lim LH, Loh YH, Wang B, Ng HH, Robson P (2005) Transcriptional regulation of nanog by OCT4 and SOX2. J Biol Chem 280:24731–24737
75. Schatz PJ (1993) Use of peptide libraries to map the substrate specificity of a peptide-modifying enzyme: a 13 residue consensus peptide specifies biotinylation in Escherichia coli. Biotechnology (N Y) 11:1138–1143
76. Silva J, Chambers I, Pollard S, Smith A (2006) Nanog promotes transfer of pluripotency after cell fusion. Nature 441:997–1001

77. Silva J, Nichols J, Theunissen TW, Guo G, van Oosten AL, Barrandon O, Wray J, Yamanaka S, Chambers I, Smith A (2009) Nanog is the gateway to the pluripotent ground state. Cell 138:722–737
78. Struhl K (2007) Transcriptional noise and the fidelity of initiation by RNA polymerase II. Nat Struct Mol Biol 14:103–105
79. Sun C, Nakatake Y, Akagi T, Ura H, Matsuda T, Nishiyama A, Koide H, Ko MS, Niwa H, Yokota T (2009) Dax1 binds to Oct3/4 and inhibits its transcriptional activity in embryonic stem cells. Mol Cell Biol 29:4574–4583
80. Suzuki A, Raya A, Kawakami Y, Morita M, Matsui T, Nakashima K, Gage FH, Rodriguez-Esteban C, Izpisua Belmonte JC (2006) Nanog binds to Smad1 and blocks bone morphogenetic protein-induced differentiation of embryonic stem cells. Proc Natl Acad Sci USA 103:10294–10299
81. Takahashi K, Tanabe K, Ohnuki M, Narita M, Ichisaka T, Tomoda K, Yamanaka S (2007) Induction of pluripotent stem cells from adult human fibroblasts by defined factors. Cell 131:861–872
82. Takahashi K, Yamanaka S (2006) Induction of pluripotent stem cells from mouse embryonic and adult fibroblast cultures by defined factors. Cell 126:663–676
83. Tanaka TS (2009) Transcriptional heterogeneity in mouse embryonic stem cells. Reprod Fertil Dev 21:67–75
84. Tang F, Barbacioru C, Bao S, Lee C, Nordman E, Wang X, Lao K, Surani MA (2010) Tracing the derivation of embryonic stem cells from the inner cell mass by single-cell RNA-Seq analysis. Cell Stem Cell 6:468–478
85. Torres J, Watt FM (2008) Nanog maintains pluripotency of mouse embryonic stem cells by inhibiting NFkappaB and cooperating with Stat3. Nat Cell Biol 10:194–201
86. Turner BM (2008) Open chromatin and hypertranscription in embryonic stem cells. Cell Stem Cell 2:408–410
87. Tutter AV, Kowalski MP, Baltus GA, Iourgenko V, Labow M, Li E, Kadam S (2009) Role for Med12 in regulation of Nanog and Nanog target genes. J Biol Chem 284:3709–3718
88. van den Berg DL, Snoek T, Mullin NP, Yates A, Bezstarosti K, Demmers J, Chambers I, Poot RA (2010) An Oct4-centered protein interaction network in embryonic stem cells. Cell Stem Cell 6:369–381
89. Wang J, Cantor AB, Orkin SH (2009) Tandem affinity purification of protein complexes in mouse embryonic stem cells using in vivo biotinylation. Curr Protoc Stem Cell Biol Chapter 1:Unit1B.5
90. Wang J, Levasseur DN, Orkin SH (2008) Requirement of Nanog dimerization for stem cell self-renewal and pluripotency. Proc Natl Acad Sci USA 105:6326–6331
91. Wang J, Rao S, Chu J, Shen X, Levasseur DN, Theunissen TW, Orkin SH (2006) A protein interaction network for pluripotency of embryonic stem cells. Nature 444:364–368
92. Wang J, Trowbridge JJ, Rao S, and Orkin SH (2008) Proteomic studies of stem cells. Stembook 1.4.1 (edited by Bernstein BE. and Lemischka I.), published online in stembook. org. StemBook [Internet]. Cambridge (MA): Harvard Stem Cell Institute; 2008-. PMID: 20614589.
93. Wang ZX, Teh CH, Kueh JL, Lufkin T, Robson P, Stanton LW (2007) Oct4 and Sox2 directly regulate expression of another pluripotency transcription factor, Zfp206, in embryonic stem cells. J Biol Chem 282:12822–12830
94. Waugh DS (2005) Making the most of affinity tags. Trends Biotechnol 23:316–320
95. Wu Q, Chen X, Zhang J, Loh YH, Low TY, Zhang W, Zhang W, Sze SK, Lim B, Ng HH (2006) Sall4 interacts with Nanog and co-occupies Nanog genomic sites in embryonic stem cells. J Biol Chem 281:24090–24094
96. Xu H, Schaniel C, Lemischka IR, Ma'ayan A (2010) Toward a complete in silico, multi-layered embryonic stem cell regulatory network. Wiley Interdiscip Rev Syst Biol Med 2:708–733
97. Xu X, Song Y, Li Y, Chang J, Zhang H, An L (2010) The tandem affinity purification method: an efficient system for protein complex purification and protein interaction identification. Protein Expr Purif 72:149–156

98. Yang J, Chai L, Fowles TC, Alipio Z, Xu D, Fink LM, Ward DC, Ma Y (2008) Genome-wide analysis reveals Sall4 to be a major regulator of pluripotency in murine-embryonic stem cells. Proc Natl Acad Sci USA 105:19756–19761
99. Ying QL, Wray J, Nichols J, Batlle-Morera L, Doble B, Woodgett J, Cohen P, Smith A (2008) The ground state of embryonic stem cell self-renewal. Nature 453:519–523
100. Yu J, Vodyanik MA, Smuga-Otto K, Antosiewicz-Bourget J, Frane JL, Tian S, Nie J, Jonsdottir GA, Ruotti V, Stewart R et al (2007) Induced pluripotent stem cell lines derived from human somatic cells. Science (New York, NY) 318:1917–1920
101. Yuan P, Han J, Guo G, Orlov YL, Huss M, Loh YH, Yaw LP, Robson P, Lim B, Ng HH (2009) Eset partners with Oct4 to restrict extraembryonic trophoblast lineage potential in embryonic stem cells. Genes Dev 23:2507–2520
102. Zhang J, Tam WL, Tong GQ, Wu Q, Chan HY, Soh BS, Lou Y, Yang J, Ma Y, Chai L et al (2006) Sall4 modulates embryonic stem cell pluripotency and early embryonic development by the transcriptional regulation of Pou5f1. Nat Cell Biol 8:1114–1123
103. Zhang X, Odom DT, Koo SH, Conkright MD, Canettieri G, Best J, Chen H, Jenner R, Herbolsheimer E, Jacobsen E et al (2005) Genome-wide analysis of cAMP-response element binding protein occupancy, phosphorylation, and target gene activation in human tissues. Proc Natl Acad Sci USA 102:4459–4464
104. Zieske LR (2006) A perspective on the use of iTRAQ reagent technology for protein complex and profiling studies. J Exp Bot 57:1501–1508

Chapter 7
Omics Era in Stem Cell Research: Data Integration of Multi-regulatory Layers

Ana Sevilla, Benjamin Trinite, and Ihor R. Lemischka

Abstract Embryonic stem cells have two unique properties: self-renewal and pluripotency (Chen and Daley, Hum Mol Genet 17:R23–R27, 2008; Jaenisch and Young, Cell 132:567–582, 2008). Self-renewal allows stem cells to generate unlimited copies of them, whereas pluripotency is the capacity to differentiate into any tissue. These two properties make embryonic stem cell research exciting for potential use in therapeutic applications (Murry and Keller, Cell 132:661–680, 2008). The transition of embryonic stem cells, from an uncommitted state to a differentiated state, involves the global execution of specific programs requiring changes in the various intra-cellular regulatory layers involving the transcriptome, the proteome, the epigenome and the methylome. During these changes, pluripotency factors are silenced and lineage-specific programs are activated in an orchestrated fashion. In this chapter we review recent efforts of studying the different regulatory layers involved in embryonic stem cells early differentiation. We describe in detail different techniques used for analysing of each regulatory layer. These techniques involve the handling of large amount of data and require the use of powerful informatics tools.

Keywords Bioinformatics tools • Epigenome • Methylome • Proteome • Regulatory layers • Stem cells systems biology • Transcriptome

A. Sevilla (✉) • I.R. Lemischka
Developmental and Regenerative Biology, Black Family Stem Cell Institute, Mount Sinai School of Medicine, New York, NY, USA
e-mail: ana.sevilla@mssm.edu

B. Trinite
Department of Basic Science, New York University College of Dentistry, New York, NY, USA

A. Ma'ayan and Ben D. MacArthur (eds.), *New Frontiers of Network Analysis in Systems Biology*, DOI 10.1007/978-94-007-4330-4_7,

7.1 Introduction

The importance of the regulatory control of stem cell fate is widely acknowledged [1–3], and a multitude of studies, both low-throughput and genome-wide scale, have aimed at understanding the regulatory processes occurring at the different compartments of the cell. These different regulatory processes include the transcriptome and microRNAs, the proteome, the epigenome which involves histone modification marks and the methylome (Fig. 7.1).

With the advent of high-throughput methods – including RNA and microRNA sequencing (RNA-seq and miRNA-seq), high throughput protein studies (SILAC, iTRAQ), and chromatin immunoprecipitation followed by sequencing (ChIP-seq), we can now have a more comprehensive view of each regulatory layer.

The significant advances in genome wide sequencing techniques and protein-protein interaction profiling have identified distinguishing characteristics for individual cell types [4–6]. Nevertheless, understanding how the regulatory layers work together to produce desired cell fate specification remains elusive. Additionally, the presence of many feedback regulatory loops among the factors that control stem cell differentiation makes their study even more complex. Therefore, there is a need to perform rigorous system dynamic studies in order to integrate several high-throughput techniques to understand the logic of the complex biological events during early differentiation of embryonic stem cells.

In this chapter, we will first introduce briefly high-throughput approaches applied to study embryonic stem cells. We will then describe few examples for each regulatory layer, and finally discuss strategies that allow the integration of the data from multiple regulatory layers.

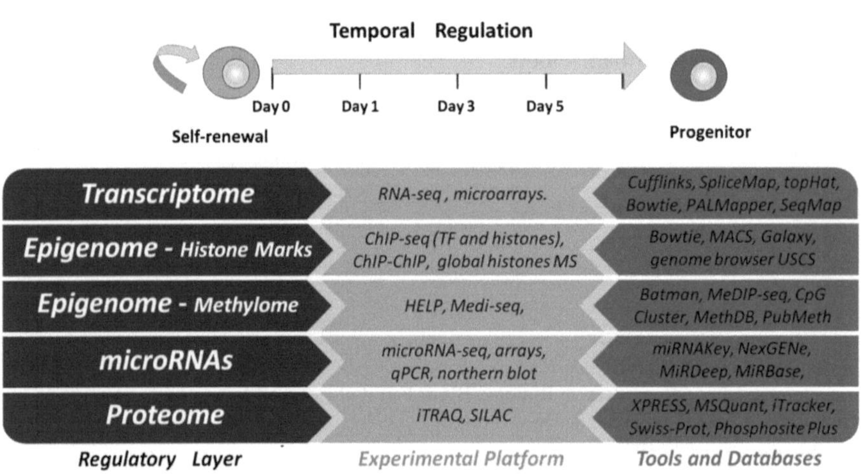

Fig. 7.1 Schematic view of dynamic system-level studies: representation of several regulatory layers, their experimental platform and several tools and databases for their study

7.2 Profiling the Transcriptome

The transcriptome is a quantitative view of the complete set of messenger RNAs (mRNAs) in a cell at a given moment for a specific developmental stage or physiological condition. Controlling the timing, location and level of expression is essential for the regulation of any biological processes, from cellular responses to external cues, to the patterning of a complete organism.

Transcriptomics aim to catalog all species of transcripts, including mRNAs, non-coding RNAs and small RNAs; to determine the transcriptional architecture of genes, mapping start sites, 5′ and 3′ ends, the possible existence of splicing patterns and other post-transcriptional RNA modifications; and to quantify the changing expression levels of each transcript during development and under different stimuli. In addition, RNA-seq can identify transcripts arising from gene fusion events which are typically found in cancer [7] as well as novel classes of non-coding RNAs (ncRNAs).

For a long time, the most established method to globally profile mRNA levels on a genome wide scale has been via microarray technology. Recently, the development of high-throughput sequencing methods provides a new method for quantifying transcriptomes with greater accuracy and potentially soon lower costs. This method, termed RNA-sequencing (RNA-seq), has clear advantages over microarrays. RNA-seq was applied to profile *Saccharomyces cerevisiae*, *Schizosaccharomyces pombe*, *Arabidopsis thaliana*, mouse and human cells [8–13]. More specifically, in the stem cell field, Wei et al. [14] presented a detailed genomic profile of Embryonic Stem Cells (ESC) lines in two different species. They compared the transcriptomes of one murine ESC line and two different human ESC lines with the method of Massive Parallel Signature Sequencing (MPSS).

As RNA-seq becomes a more affordable, the transcriptional profiling of the whole genome is likely to become the platform of choice. The ability of the sequencing platform to interrogate the entire transcriptome in contrast with tiling microarrays provides new insights into the levels of transcriptional complexity. RNA-seq provides quantitative measures of splicing variants and single nucleotide polymorphisms (SNPs) for allele specific expression without any prior knowledge. However, this new level of detail requires careful statistical analysis to deliver the promise of this new technology.

In principle, any high-throughput sequencing technology [15] can be used for RNA-seq. Those include the Illumina IG [8, 9, 11, 12, 16] Applied Biosystems SOLiD [10] and Roche 454 Life Science [17–19]. RNA-seq can detect alternative splicing variants using paired-end, relatively short reads (on the Illumina and Applied Biosystems platforms) or longer reads (using the Roche platform).

As RNA-seq is quantitative, it can be used to determine RNA expression levels more accurately than microarrays. In principle, it is possible to determine the absolute quantity of every RNA molecule in a cell population, and directly compare results between experiments. Several methods have been used for transcript quantification [12, 13]. However, some care should be applied. For example, the Oshlack group have observed that the length of the transcripts bias the identification of differentially expressed genes. This is a common problem with RNA-seq technologies [20].

7.2.1 Tools for RNA-seq Data Analysis

Many tools are rapidly emerging for the RNA-seq data analysis. For example, Scripture [21] is a method for transcriptome reconstruction that relies solely on RNA-seq reads. Cufflinks [22] is software that assembles transcripts, estimates their abundances, and tests for differential expression. SpliceMap [23] is a de novo splice junction discovery tool. It offers high sensitivity and support for arbitrarily long RNA-seq read lengths. TopHat [24] is a fast splice junction mapper for RNA-seq reads. It aligns RNA-seq reads to mammalian-sized genomes using the ultra high-throughput short read aligner Bowtie [25], and then analyzes the mapping results to identify splice junctions between exons. PALMapper [26] is a combination of the spliced alignment method QPALMA [27] with the short read alignment tool GenomeMapper [28]. The resulting method, called PALMapper, efficiently computes both spliced and unspliced alignments at high accuracy while taking advantage of single base level quality information and splice site predictions. SeqMap [29] is a tool for mapping millions of short sequences to the genome. Bioconductor is an open source package for the analysis and comprehension of genomic data containing packages for RNA-sequencing analysis. BWA [30], is a fast light-weighted tool that aligns relatively short sequences to a sequence database, such as the human reference genome. The combination of several tools will facilitate a rigorous RNA-seq data analysis.

7.3 Profiling microRNAs

microRNAs (miRNAs) are short ribonucleic acid (RNA) molecules, on average only 22 nucleotides long. The human genome may encode over different 1,000 miRNAs [31] which may target about 60% of all mammalian genes [32]. MiRNAs are post-transcriptional regulators that bind partially to complementary sequences in the three prime untranslated regions ($3'$ UTRs) of target messenger RNA transcripts (mRNAs), usually resulting in gene silencing [33, 34]. Each miRNA may repress hundreds of mRNAs [35, 36]. Consequently miRNAs are important regulators of gene expression.

There are different methods to quantify microRNA expression; by RT-qPCR, Northern, microarrays (Exiqon) and through sequencing platforms. Protocols for sequencing microRNAs involve several steps; first isolation of the mature microRNA fraction and then cloning these small regulatory molecules with the adaptors for the sequencer [37, 38].

Recently, Marson et al. [39] integrated ChIP-seq data for important pluripotent transcription factors of mESC and ChIP-seq for the histone modification H3K4me3 to demonstrate that Oct4, Sox2, Nanog and Tcf3 play an important role in regulating the expression of several microRNAs at the promoter level. Some of these miRNAs turn to be highly expressed in ES cells, while other specific groups of miRNAs are only expressed in differentiated cells. In relation to this observation, Xu and Kosik [40], describe that cell-type-specific expression signatures of microRNAs in mouse and human can successfully allowed to distinguish different cell states and fates.

7.3.1 Tools for miRNA-seq Data Analysis

Several software for miRNA data analysis have been developed recently, for example, miRNAkey is a software for microRNA deep sequencing analysis [41]. NextGENe is software for small RNA analysis. MiRDeep [42] is a package developed to discover active known or novel miRNAs from deep sequencing data (Solexa/Illumina, 454), miRanalyser [43] is a web server tool for the analysis of deep-sequencing experiments for small RNAs.

Some of the primary online resources for miRNA sequences, target prediction, and other annotations are Miranda [44], TargetScan [45], PicTar [46], DIANA-microT [47, 48], rna22, MirTarget2 [49], NBmirTar [50] and PITA [51] a successful algorithm that considers secondary structure.

7.4 Profiling the Epigenome

Genome-wide mapping of protein–DNA interactions and epigenetic marks is essential for a full understanding of transcriptional regulation [52]. The combination of nucleosome positioning and dynamic modification of DNA and histones has a key role in gene regulation [53–55] as it guides development and differentiation. Moreover, chromatin state influences transcription directly, conditioning the packaging of DNA to allow or prevent access to DNA-binding proteins. DNA methylation and covalent modifications of histone proteins have also been broadly defined as epigenetic modifications [56, 57] and are important for transcriptional control [58–60]. The program of gene expression is therefore dependent on the epigenome. Chromatin structure plays also a role in other genomic activities such as development and cancer. For example, it has been shown that changes in chromatin structure are important in silencing certain genes involved in cancer, and histone deacetylase inhibitors have demonstrated anticancer effects [61]. Finally, during the development, the regulation of chromatin structure has emerged as a key mechanism to modulate developmental gene expression patterns. Recent global analyses have revealed key differences in the chromatin landscape of pluripotent embryonic stem (ES) cells by comparison with lineage-committed cells, suggesting that chromatin states may be linked to cell fate [4]. All these advances suggest that this interplay between chromatin and transcription is dynamic and more complex than previously appreciated [62], and there is a growing recognition that systematic profiling of the epigenome in multiple cell types and stages may be needed for understanding developmental processes and disease states [56].

All these epigenomic studies started with ChIP-chip technology that combined chromatin immunoprecipitation with microarray technology. At the present, ChIP-PET and ChIP-seq techniques are successfully being used for single read sequencing and whole genome coverage.

7.4.1 Tools for ChIP-seq Data Analysis

Some of the public available ChIP-seq software for 'peak-finding' are: FindPeaks [63], F-Seq [64], SISSRS [65], Peak Seq [66], QuEST [67], MACS [68], the ChipSeq Peak Finder used in [69], ChIPDiff [70] and CisGenome [71], which encompasses a series of tools for the different steps of the ChIP-seq analysis pipeline. False discovery rates are estimated by these tools by comparing the level of enrichment (number of tags) at given sites, with the background model used.

7.5 Profiling the Methylome

DNA methylation provides, in many eukaryotes, another layer of epigenetic control that has important implications for normal biology and disease. Bisulphite genomic sequencing is the most common technology for detection of 5-Methylcytosine, which maps sites at single base-pair resolution [72]. This method depends on the identification of, following previous incubation with sodium bisulphite, cytosines in single-stranded DNA which are deaminated to give uracil. There have been successful bisulphite DNA sequencing studies [73, 74], but this is a time-consuming and resource-intensive process. Therefore, alternative attempts to map DNA methylation on a genome-wide scale have so far relied on less direct methods. Approaches based on the sensitivity of restriction enzymes to CpG methylation within their cleavage recognition site [75] are comparatively low resolution, but they are useful when they are combined with microarrays [76, 77]. In summary, the methods developed to study DNA methylation can be grouped into several general categories: (1) Methods that employ the bisulphite conversion of cytosine residues (but not 5-mC) into uracil [72, 78, 79]. (2) Methylation-sensitive restriction endonuclease-based methods [80, 81]. (3) Enrichment methods, such as methyl-DNA immunoprecipitation, based on antibodies specific to 5-mC [82, 83]. (4) The capture of methylated DNA using proteins containing methyl-CpG-binding domains (MBD) [84–86].

Studies using high-throughput bisulphite sequencing and single-molecule-based sequencing sampling of conserved non-coding elements, transposons and other genomic features, for mouse embryonic stem cells, embryonic stem cell derived and primary neural cells showed that DNA methylation patterns are better correlated with histone methylation patterns than with the underlying genome sequence context [87]. Therefore, profiling DNA methylation across the entire genome is vital for understanding the epigenetic regulation.

7.5.1 Tools for Methylome Data Analysis

Leading tools for methylome data analysis are: Batman [88] a cross-platform algorithm bayesian tool for analyzing methylated DNA immunoprecipitated (MeDIP) profiles generated using oligonucleotide arrays (MeDIP-chip) or next-generation sequencing (MeDIP-seq). Bisulphite sequencing Data Presentation and Compilation (BDPC) [89] is a DNA methylation analysis platform for analysis and clustering primary bisulphite sequencing data. Bisulphite Sequence Mapping Program (BSMAP) [90] is a software for sort reads mapping for bisulphite sequencing reads, CpGcluster [91], CpG Island Explorer [92] and CpG Island Searcher [93] are all tools for online CpG Island identification, and finally CpG Promoter [94] is a tool for large-scale promoter mapping using CpG islands.

7.6 Profiling the Proteome and Phosphoproteome

The aim of quantitative proteomics is to obtain about the quantity of all proteins expressed in a sample [95, 96]· Proteins carry out all biological processes, and a global analysis of their regulation and molecular function is essential to better understand biological systems at the system level. In addition to transcriptional and translational regulation, cell-fate determination is also controlled by protein phosphorylation, a critical determinant of cell signalling [97, 98].

Among the current methods to study the proteome at the systems scale include: (1) Isotope-Coded Affinity Tags (ICAT) [99]; (2) Isobaric labelling: tandem mass tags (TMT) [100] and isobaric tags for relative and absolute quantization (iTRAQ) [101]; (3) Label-free quantization no label quantization [102]; (4) Metal Coded Tags (MeCATs) [103]; (5) N-Terminal Labelling [104]; (6) Stable Isotope Labelling with Amino Acids in Cell Culture (SILAC) [105].

Improvements in phosphoproteomics technology now allow sequencing of hundreds or even thousands of phosphorylation sites in a single experiment [97, 106]. However, it has rapidly become clear that a functional filter is needed to extract biological insights from such experiments. An alternative method called iTRAQ uses stable isotope labelling at the tag-level. The fragmentation of the attached tag generates a low molecular mass reporter ion that can be used to relatively quantify the peptides and the proteins from which they originated, using software such as the freely available i-Tracker [107].

Examples of high throughput phosphoproteomic studies to profile stem cells include, Brill et al. [108] who performed a large-scale multidimensional liquid chromatography (MDLC)- tandem mass spectrometry (MS/MS)-based phosphoproteomic analysis of undifferentiated hESCs and their differentiated derivatives for identification of protein phosphorylation sites in these cells. They showed several phosphoproteins involved in receptor tyrosine kinase (RTK) signalling pathways. Cellular assays corroborated this observation by showing that multiple RTKs cooperatively support the undifferentiated hESCs state. Since ESCs are a promising source for regenerative therapy, characterizating the phosphoproteome status of ESCs will enhance our understanding of the molecular mechanisms and signalling pathways that regulate their propagation and pluipotency and how external stimuli drive ES cell signalling toward differentiation.

7.6.1 Tools for Proteome and Phosphoproteome Analysis

The computational requirements for analyzing data for quantification of isotopic labelling experiments are similar for the different strategies where the major difference between the utilized labels lays in the mass shift of generate a isotopic pairs of peptides. Among the programs available for SILAC proteomics data analysis, the open source program MSQuant [109] allows quantification of SILAC data; Multi-Q [110], iTracker [107], and the "iTRAQ reporter ion counter" [111] are other freely available software, which import pre-processed MS/MS data from Sequest or Mascot [112]. Libra [113] is another freely available program for iTRAQ quantification and is integrated as a module in the Trans Proteomic Pipeline (TPP), which is an open source platform for the visual assessment and statistical evaluation of MS/MS peptide assignments. The commercial programs ProQuant and ProteinPilot distributed by Applied Biosystems offer more sophisticated software solutions. An alternative to MS-based proteomics is antibody arrays. For example, Kinexus and Full Moon Bioscience have specialized methods for analyzing the phosphoproteome using antibody-based microarrays, kinase substrate, and inhibitor profiling. Other companies like R&D, also offer antibody based microarrays of pathways to be able to study specific biological networks under different experimental conditions.

7.7 Mapping the Interactome

Interactions, both physical and functional, are critical for advancing functional genomics. ChIP–chip/ ChIP-seq are able to provide genome-scale information on DNA–protein interactions, and high-throughput sequencing of RNAs isolated by crosslinking and immunoprecipitation (HITS-CLIP, also known as CLIP-seq) is emerging as an important method for understanding RNA–protein interactions [114]. High-throughput protein–protein interaction networks have been largely uncovered by the two-hybrid system [115]. However, immunoprecipitation followed by mass spectrometry is becoming more widely accepted as the method of choice to profile protein-protein interactions [116]. Recently, the Nanog interactome was presented by Wang et al. [6] and similar approaches have mapped the Oct4 interactome [117].

7.8 Bioinformatics Challenges

The broad array of data set types collected from different experimental platforms provides huge opportunities and challenges for investigators to address some important biological questions related to fundamental mechanisms of genome function. However, with high-throughput sequencing technologies such as, RNA-seq, microRNA-seq and ChIP-seq, there are several informatics challenges, including the development of efficient methods to store and process large amounts of data.

Once high-quality reads have been filtered, the first step in the data analysis of deep sequencing data analysis pipeline is to align the short reads from RNA-seq, microRNA-seq or ChIP-seq to the reference genome. There are several programs for mapping reads to the genome, including ELAND [118], SOAP [119], MAQ [120], Bowtie [25] and RMAP [121]. To profile the enriched binding regions of transcription factors or specific histone modification marks there exist several programs: CisGenome [71] is an open access software that is compatible with several operating systems, file-conversion tools, and tools to call peaks of ChIP enrichment and to perform motif analysis. The Model Based Analysis for ChIP-seq (MACS) [122] is a useful package for ChIP-seq data analysis. A commonly used method of data integration is to perform intersection analysis on enriched regions from different experiments. For example, Chen et al. mapped a collection of 13 transcription factors using ChIP-seq technology in mouse embryonic stem cells, then used a custom peak-finder to call regions of enrichment and observed significant co-binding of transcription factors [123].

Although intersection analysis of the same output format of different experimental approaches is straightforward to perform, special attention must be paid to the underlying assumptions of different data sets coming from different platforms with different output formats. This issue is often addressed by sophisticated methods for normalization.

7.9 Data Integration

The genome serves as a platform where high-throughput data, from different regulatory layers, can be anchored. Clustering can be proposed as a first stage of data integration classifying genomic loci into groups with shared similar patterns.

For instance, clustering of RNA expression shows genes that are co-expressed [124], clustering of histone modifications disclose loci that share similar chromatin structure [125, 126], protein–protein interaction clustering depicts proteins in the same complex [127], and genetic interaction clustering reveals members of the same or similar pathways [128].

The next level of data integration is to connect different regulatory layers "inside another". One of the most representative studies of multi regulatory data integration in stem cells has been the systems-level dynamic analyses of fate change in murine stem cells, conducted by Lu et al. [129]. This study challenged the effort in data integration from different regulatory layers where global changes in histone acetylation, chromatin bound RNA polymerase II, messenger RNA (mRNA), and nuclear protein levels were measured in cells induced to differentiation over 5 days after downregulation of Nanog a key pluripotency regulator [130, 131].

Other good examples of next generation data sequencing integration studies have been for example the studies of chromatin signature, in the context of DNA methylation [16, 132, 133].

Coupling histone modification data to transcriptomic data can also be valuable for the annotation of the recently discovered non coding RNAs (ncRNAs). Young and colleagues [39] identified microRNA (miRNA) transcription start sites by mapping the promoter-specific modification histone 3 lysine 4 trimethylation (H3K4me3) and comparing regions outside known promoters with annotated miRNAs, conserved regions, CpG islands and histone modifications (H3K36me3 and H3K79me2) that are associated with transcription elongation. Similarly, Rinn and colleagues [134] mapped the location of thousands of large intervening non-coding (lincRNAs) by integrating these same chromatin modifications with RNA-seq data for expressed ncRNAs. It is now thought that many of these lincRNAs can influence histone modification or chromatin structure or subsequent methylation of DNA [135–137].

Data integration from three different regulatory layers: epigenomics, genomics and transcriptomics can also supply a comprehensive view of transcription-coupled RNA processing. For instance, ChIP-seq studies of the methylation mark H3K36me3 showed evidence of good correlation between the levels of this methylation mark and exon expression [126, 138–142].

To move forward in the era of next generation sequencing (NGS), we need user friendly tools for the bench scientist to analyze their own data in an efficient and straightforward manner. A tool that reaches these qualities is Galaxy [143, 144], which is an open-source web browser, where a core group of programmers maintain the browser code and interact actively with the community in order to update and develop new tools according to the emerging needs.

Another important challenge is how to visualize all these data. The Genome Browser, developed by the University of California-Santa Cruz (UCSC), constitutes a useful platform for alignment of data sets in the genome. By comparing new information generated with previous known annotations can provide an easy bridge to hypothesis generation. In the past, when genomics consisted only of global gene expression analysis, annotation libraries such as Gene Ontology [145] and Gene Set Enrichment Analysis (GSEA) [146] were developed to provide an easy way to assess the biological significance of gene hits. In this next generation of genomic data analysis, with the recent discovery of non-coding RNA (ncRNAs) and disease associated single-nucleotide polymorphism (SNPs), it is more evident that the need for additional tools integrating different data sets for are required. Ensemble of loci that share factor binding, epigenetic modifications or disease association may provide efficient ways to formulate hypotheses regarding function outside of coding regions.

7.10 Future Directions

It is evident that we still are at the early stages of data processing and normalization. For example, RNA-seq data need more standardize guidelines. It remains unclear how RNA-seq data from platforms that sequence short tags will be normalized against data from longer read platforms, or how relative levels of expression can be compared, especially if there is variation in the number of reads sequenced. To address these questions in more detail, it will be important to revise data normalization and generate certain standards in the field. Therefore, we will be able to compare RNA-seq experiments among different groups and experimental settings, as it has been done for years with the microarray platform. In the case of ChIP-seq studies, experimental variation or antibody differences (for ChIP-seq or HITS-CLIP) have been shown to provide varying degrees of enrichment. Therefore, all data sets used in a comparative analysis should first be normalized with the same standards. This applies to samples from different research groups, as well as to samples from within a data set. Normalization is therefore imperative in experiments that examine time points of a differentiation process or stages of disease progression in which the changes may be subtle between stages [147].

Data integration itself is not the end: the final goal is to formulate new rational hypotheses based on evidence from the data analyzed and help to test them in a low throughput scale in functional experiments.

The future of genomic and proteomic technologies hold great expectations, but for genomic and proteomics data to achieve their potential, computational integration must link together all of the information that is generated. For this purpose, the Encyclopaedia of DNA Elements (ENCODE) consortium is performing an exemplar role [148]. It is an international collaboration of research groups with the goal of building a comprehensive part list of functional elements in the human genome. In this platform, data-set is presented in a user friendly manner and each gene entry is linked to the UCSC-Genome Browser for visualizing all genomic and epigenomic information. This type of integration of knowledge at different regulatory levels will give to the scientist a broad picture for greater understanding of all the biological systems.

References

1. Chen L, Daley GQ (2008) Molecular basis of pluripotency. Hum Mol Genet 17(R1): R23–R27
2. Jaenisch R, Young R (2008) Stem cells, the molecular circuitry of pluripotency and nuclear reprogramming. Cell 132(4):567–582
3. Murry CE, Keller G (2008) Differentiation of embryonic stem cells to clinically relevant populations: lessons from embryonic development. Cell 132(4):661–680
4. Mikkelsen TS, Ku M, Jaffe DB, Issac B, Lieberman E, Giannoukos G, Alvarez P, Brockman W, Kim TK, Koche RP et al (2007) Genome-wide maps of chromatin state in pluripotent and lineage-committed cells. Nature 448(7153):553–560
5. Graumann J, Hubner NC, Kim JB, Ko K, Moser M, Kumar C, Cox J, Scholer H, Mann M (2008) Stable isotope labeling by amino acids in cell culture (SILAC) and proteome quantitation of mouse embryonic stem cells to a depth of 5,111 proteins. Mol Cell Proteomics 7 (4):672–683
6. Wang J, Rao S, Chu J, Shen X, Levasseur DN, Theunissen TW, Orkin SH (2006) A protein interaction network for pluripotency of embryonic stem cells. Nature 444(7117):364–368
7. Maher CA, Kumar-Sinha C, Cao X, Kalyana-Sundaram S, Han B, Jing X, Sam L, Barrette T, Palanisamy N, Chinnaiyan AM (2009) Transcriptome sequencing to detect gene fusions in cancer. Nature 458(7234):97–101
8. Wilhelm BT, Marguerat S, Watt S, Schubert F, Wood V, Goodhead I, Penkett CJ, Rogers J, Bahler J (2008) Dynamic repertoire of a eukaryotic transcriptome surveyed at single-nucleotide resolution. Nature 453(7199):1239–1243
9. Mortazavi A, Williams BA, McCue K, Schaeffer L, Wold B (2008) Mapping and quantifying mammalian transcriptomes by RNA-Seq. Nat Methods 5(7):621–628
10. Cloonan N, Forrest AR, Kolle G, Gardiner BB, Faulkner GJ, Brown MK, Taylor DF, Steptoe AL, Wani S, Bethel G et al (2008) Stem cell transcriptome profiling via massive-scale mRNA sequencing. Nat Methods 5(7):613–619
11. Marioni JC, Mason CE, Mane SM, Stephens M, Gilad Y (2008) RNA-seq: an assessment of technical reproducibility and comparison with gene expression arrays. Genome Res 18 (9):1509–1517
12. Morin R, Bainbridge M, Fejes A, Hirst M, Krzywinski M, Pugh T, McDonald H, Varhol R, Jones S, Marra M (2008) Profiling the HeLa S3 transcriptome using randomly primed cDNA and massively parallel short-read sequencing. Biotechniques 45(1):81–94
13. Nagalakshmi U, Wang Z, Waern K, Shou C, Raha D, Gerstein M, Snyder M (2008) The transcriptional landscape of the yeast genome defined by RNA sequencing. Science 320 (5881):1344–1349
14. Wei CL, Miura T, Robson P, Lim SK, Xu XQ, Lee MY, Gupta S, Stanton L, Luo Y, Schmitt J et al (2005) Transcriptome profiling of human and murine ESCs identifies divergent paths required to maintain the stem cell state. Stem Cells 23(2):166–185
15. Holt RA, Jones SJ (2008) The new paradigm of flow cell sequencing. Genome Res 18 (6):839–846
16. Lister R, O'Malley RC, Tonti-Filippini J, Gregory BD, Berry CC, Millar AH, Ecker JR (2008) Highly integrated single-base resolution maps of the epigenome in Arabidopsis. Cell 133(3):523–536
17. Barbazuk WB, Emrich SJ, Chen HD, Li L, Schnable PS (2007) SNP discovery via 454 transcriptome sequencing. Plant J 51(5):910–918
18. Vera JC, Wheat CW, Fescemyer HW, Frilander MJ, Crawford DL, Hanski I, Marden JH (2008) Rapid transcriptome characterization for a nonmodel organism using 454 pyrosequencing. Mol Ecol 17(7):1636–1647
19. Emrich SJ, Barbazuk WB, Li L, Schnable PS (2007) Gene discovery and annotation using LCM-454 transcriptome sequencing. Genome Res 17(1):69–73

20. Oshlack A, Wakefield MJ (2009) Transcript length bias in RNA-seq data confounds systems biology. Biol Direct 4:14
21. Guttman M, Garber M, Levin JZ, Donaghey J, Robinson J, Adiconis X, Fan L, Koziol MJ, Gnirke A, Nusbaum C et al (2010) Ab initio reconstruction of cell type-specific transcriptomes in mouse reveals the conserved multi-exonic structure of lincRNAs. Nat Biotechnol 28(5):503–510
22. Roberts A, Trapnell C, Donaghey J, Rinn JL, Pachter L (2010) Improving RNA-seq expression estimates by correcting for fragment bias. Genome Biol 12(3):R22
23. Au KF, Jiang H, Lin L, Xing Y, Wong WH (2010) Detection of splice junctions from paired-end RNA-seq data by SpliceMap. Nucleic Acids Res 38(14):4570–4578
24. Trapnell C, Pachter L, Salzberg SL (2009) TopHat: discovering splice junctions with RNA-Seq. Bioinformatics 25(9):1105–1111
25. Langmead B, Trapnell C, Pop M, Salzberg SL (2009) Ultrafast and memory-efficient alignment of short DNA sequences to the human genome. Genome Biol 10(3):R25
26. Jean G, Kahles A, Sreedharan VT, De Bona F, Ratsch G (2010) RNA-seq read alignments with PALMapper. Curr Protoc Bioinform Chapter 11:Unit 11.16
27. De Bona F, Ossowski S, Schneeberger K, Ratsch G (2008) Optimal spliced alignments of short sequence reads. Bioinformatics 24(16):i174–i180
28. Schneeberger K, Hagmann J, Ossowski S, Warthmann N, Gesing S, Kohlbacher O, Weigel D (2009) Simultaneous alignment of short reads against multiple genomes. Genome Biol 10(9):R98
29. Jiang H, Wong WH (2008) SeqMap: mapping massive amount of oligonucleotides to the genome. Bioinformatics 24(20):2395–2396
30. Li H, Durbin R (2009) Fast and accurate short read alignment with Burrows-Wheeler transform. Bioinformatics 25(14):1754–1760
31. Bentwich I, Avniel A, Karov Y, Aharonov R, Gilad S, Barad O, Barzilai A, Einat P, Einav U, Meiri E et al (2005) Identification of hundreds of conserved and nonconserved human microRNAs. Nat Genet 37(7):766–770
32. Friedman RC, Farh KK, Burge CB, Bartel DP (2009) Most mammalian mRNAs are conserved targets of microRNAs. Genome Res 19(1):92–105
33. Bartel DP (2009) MicroRNAs: target recognition and regulatory functions. Cell 136 (2):215–233
34. Bartel DP (2004) MicroRNAs: genomics, biogenesis, mechanism, and function. Cell 116 (2):281 297
35. Brennecke J, Stark A, Russell RB, Cohen SM (2005) Principles of microRNA-target recognition. PLoS Biol 3(3):e85
36. Lim LP, Lau NC, Garrett-Engele P, Grimson A, Schelter JM, Castle J, Bartel DP, Linsley PS, Johnson JM (2005) Microarray analysis shows that some microRNAs downregulate large numbers of target mRNAs. Nature 433(7027):769–773
37. Pfeffer S, Lagos-Quintana M, Tuschl T (2005) Cloning of small RNA molecules. Curr Protoc Mol Biol Chapter 26:Unit 26.24
38. Lau NC, Lim LP, Weinstein EG, Bartel DP (2001) An abundant class of tiny RNAs with probable regulatory roles in Caenorhabditis elegans. Science 294(5543):858–862
39. Marson A, Levine SS, Cole MF, Frampton GM, Brambrink T, Johnstone S, Guenther MG, Johnston WK, Wernig M, Newman J et al (2008) Connecting microRNA genes to the core transcriptional regulatory circuitry of embryonic stem cells. Cell 134(3):521–533
40. Xu N, Papagiannakopoulos T, Pan G, Thomson JA, Kosik KS (2009) MicroRNA-145 regulates OCT4, SOX2, and KLF4 and represses pluripotency in human embryonic stem cells. Cell 137(4):647–658
41. Ronen R, Gan I, Modai S, Sukacheov A, Dror G, Halperin E, Shomron N (2010) miRNAkey: a software for microRNA deep sequencing analysis. Bioinformatics 26(20):2615–2616

42. Friedlander MR, Chen W, Adamidi C, Maaskola J, Einspanier R, Knespel S, Rajewsky N (2008) Discovering microRNAs from deep sequencing data using miRDeep. Nat Biotechnol 26(4):407–415
43. Hackenberg M, Sturm M, Langenberger D, Falcon-Perez JM, Aransay AM (2009) miRanalyzer: a microRNA detection and analysis tool for next-generation sequencing experiments. Nucleic Acids Res 37(Web Server issue):W68–W76
44. Betel D, Wilson M, Gabow A, Marks DS, Sander C (2008) The microRNA.org resource: targets and expression. Nucleic Acids Res 36(Database issue):D149–D153
45. Lewis BP, Burge CB, Bartel DP (2005) Conserved seed pairing, often flanked by adenosines, indicates that thousands of human genes are microRNA targets. Cell 120(1):15–20
46. Krek A, Grun D, Poy MN, Wolf R, Rosenberg L, Epstein EJ, MacMenamin P, da Piedade I, Gunsalus KC, Stoffel M et al (2005) Combinatorial microRNA target predictions. Nat Genet 37(5):495–500
47. Maragkakis M, Alexiou P, Papadopoulos GL, Reczko M, Dalamagas T, Giannopoulos G, Goumas G, Koukis E, Kourtis K, Simossis VA et al (2009) Accurate microRNA target prediction correlates with protein repression levels. BMC Bioinformatics 10:295
48. Maragkakis M, Reczko M, Simossis VA, Alexiou P, Papadopoulos GL, Dalamagas T, Giannopoulos G, Goumas G, Koukis E, Kourtis K et al (2009) DIANA-microT web server: elucidating microRNA functions through target prediction. Nucleic Acids Res 37 (Web Server issue):W273–W276
49. Wang X, El Naqa IM (2008) Prediction of both conserved and nonconserved microRNA targets in animals. Bioinformatics 24(3):325–332
50. Yousef M, Jung S, Kossenkov AV, Showe LC, Showe MK (2007) Naive Bayes for microRNA target predictions–machine learning for microRNA targets. Bioinformatics 23 (22):2987–2992
51. Kertesz M, Iovino N, Unnerstall U, Gaul U, Segal E (2007) The role of site accessibility in microRNA target recognition. Nat Genet 39(10):1278–1284
52. Farnham PJ (2009) Insights from genomic profiling of transcription factors. Nat Rev Genet 10 (9):605–616
53. Jiang C, Pugh BF (2009) Nucleosome positioning and gene regulation: advances through genomics. Nat Rev Genet 10(3):161–172
54. Henikoff S (2008) Nucleosome destabilization in the epigenetic regulation of gene expression. Nat Rev Genet 9(1):15–26
55. Li B, Carey M, Workman JL (2007) The role of chromatin during transcription. Cell 128(4):707–719
56. Bernstein BE, Meissner A, Lander ES (2007) The mammalian epigenome. Cell 128(4):669–681
57. Goldberg AD, Allis CD, Bernstein E (2007) Epigenetics: a landscape takes shape. Cell 128(4):635–638
58. Jones PA, Baylin SB (2007) The epigenomics of cancer. Cell 128(4):683–692
59. Kouzarides T (2007) Chromatin modifications and their function. Cell 128(4):693–705
60. Li E (2002) Chromatin modification and epigenetic reprogramming in mammalian development. Nat Rev Genet 3(9):662–673
61. Baylin SB (2005) DNA methylation and gene silencing in cancer. Nat Clin Pract Oncol 2 (Suppl 1):S4–S11
62. Berger SL (2007) The complex language of chromatin regulation during transcription. Nature 447(7143):407–412
63. Fejes AP, Robertson G, Bilenky M, Varhol R, Bainbridge M, Jones SJ (2008) FindPeaks 3.1: a tool for identifying areas of enrichment from massively parallel short-read sequencing technology. Bioinformatics 24(15):1729–1730
64. Boyle AP, Guinney J, Crawford GE, Furey TS (2008) F-Seq: a feature density estimator for high-throughput sequence tags. Bioinformatics 24(21):2537–2538
65. Jothi R, Cuddapah S, Barski A, Cui K, Zhao K (2008) Genome-wide identification of in vivo protein-DNA binding sites from ChIP-seq data. Nucleic Acids Res 36(16):5221–5231

66. Rozowsky J, Euskirchen G, Auerbach RK, Zhang ZD, Gibson T, Bjornson R, Carriero N, Snyder M, Gerstein MB (2009) PeakSeq enables systematic scoring of ChIP-seq experiments relative to controls. Nat Biotechnol 27(1):66–75
67. Valouev A, Johnson DS, Sundquist A, Medina C, Anton E, Batzoglou S, Myers RM, Sidow A (2008) Genome-wide analysis of transcription factor binding sites based on ChIP-seq data. Nat Methods 5(9):829–834
68. Cokus SJ, Feng S, Zhang X, Chen Z, Merriman B, Haudenschild CD, Pradhan S, Nelson SF, Pellegrini M, Jacobsen SE (2008) Shotgun bisulphite sequencing of the Arabidopsis genome reveals DNA methylation patterning. Nature 452(7184):215–219
69. Johnson DS, Mortazavi A, Myers RM, Wold B (2007) Genome-wide mapping of in vivo protein-DNA interactions. Science 316(5830):1497–1502
70. Xu H, Wei CL, Lin F, Sung WK (2008) An HMM approach to genome-wide identification of differential histone modification sites from ChIP-seq data. Bioinformatics 24(20):2344–2349
71. Ji H, Jiang H, Ma W, Johnson DS, Myers RM, Wong WH (2008) An integrated software system for analyzing ChIP-chip and ChIP-seq data. Nat Biotechnol 26(11):1293–1300
72. Frommer M, McDonald LE, Millar DS, Collis CM, Watt F, Grigg GW, Molloy PL, Paul CL (1992) A genomic sequencing protocol that yields a positive display of 5-methylcytosine residues in individual DNA strands. Proc Natl Acad Sci USA 89(5):1827–1831
73. Eckhardt F, Lewin J, Cortese R, Rakyan VK, Attwood J, Burger M, Burton J, Cox TV, Davies R, Down TA et al (2006) DNA methylation profiling of human chromosomes 6, 20 and 22. Nat Genet 38(12):1378–1385
74. Rollins RA, Haghighi F, Edwards JR, Das R, Zhang MQ, Ju J, Bestor TH (2006) Large-scale structure of genomic methylation patterns. Genome Res 16(2):157–163
75. Bird AP, Southern EM (1978) Use of restriction enzymes to study eukaryotic DNA methylation: I. The methylation pattern in ribosomal DNA from Xenopus laevis. J Mol Biol 118 (1):27–47
76. Schumacher A, Kapranov P, Kaminsky Z, Flanagan J, Assadzadeh A, Yau P, Virtanen C, Winegarden N, Cheng J, Gingeras T et al (2006) Microarray-based DNA methylation profiling: technology and applications. Nucleic Acids Res 34(2):528–542
77. Khulan B, Thompson RF, Ye K, Fazzari MJ, Suzuki M, Stasiek E, Figueroa ME, Glass JL, Chen Q, Montagna C et al (2006) Comparative isoschizomer profiling of cytosine methylation: the HELP assay. Genome Res 16(8):1046–1055
78. Herman JG, Graff JR, Myohanen S, Nelkin BD, Baylin SB (1996) Methylation-specific PCR: a novel PCR assay for methylation status of CpG islands. Proc Natl Acad Sci USA 93 (18):9821–9826
79. Eads CA, Danenberg KD, Kawakami K, Saltz LB, Blake C, Shibata D, Danenberg PV, Laird PW (2000) MethyLight: a high-throughput assay to measure DNA methylation. Nucleic Acids Res 28(8):E32
80. Dahl C, Guldberg P (2003) DNA methylation analysis techniques. Biogerontology 4 (4):233–250
81. Irizarry RA, Ladd-Acosta C, Carvalho B, Wu H, Brandenburg SA, Jeddeloh JA, Wen B, Feinberg AP (2008) Comprehensive high-throughput arrays for relative methylation (CHARM). Genome Res 18(5):780–790
82. Weber M, Davies JJ, Wittig D, Oakeley EJ, Haase M, Lam WL, Schubeler D (2005) Chromosome-wide and promoter-specific analyses identify sites of differential DNA methylation in normal and transformed human cells. Nat Genet 37(8):853–862
83. Acevedo LG, Bieda M, Green R, Farnham PJ (2008) Analysis of the mechanisms mediating tumor-specific changes in gene expression in human liver tumors. Cancer Res 68(8):2641–2651
84. Cross SH, Charlton JA, Nan X, Bird AP (1994) Purification of CpG islands using a methylated DNA binding column. Nat Genet 6(3):236–244
85. Klose RJ, Sarraf SA, Schmiedeberg L, McDermott SM, Stancheva I, Bird AP (2005) DNA binding selectivity of MeCP2 due to a requirement for A/T sequences adjacent to methyl-CpG. Mol Cell 19(5):667–678

86. Gebhard C, Schwarzfischer L, Pham TH, Andreesen R, Mackensen A, Rehli M (2006) Rapid and sensitive detection of CpG-methylation using methyl-binding (MB)-PCR. Nucleic Acids Res 34(11):e82
87. Meissner A, Mikkelsen TS, Gu H, Wernig M, Hanna J, Sivachenko A, Zhang X, Bernstein BE, Nusbaum C, Jaffe DB et al (2008) Genome-scale DNA methylation maps of pluripotent and differentiated cells. Nature 454(7205):766–770
88. Down TA, Rakyan VK, Turner DJ, Flicek P, Li H, Kulesha E, Graf S, Johnson N, Herrero J, Tomazou EM et al (2008) A Bayesian deconvolution strategy for immunoprecipitation-based DNA methylome analysis. Nat Biotechnol 26(7):779–785
89. Rohde C, Zhang Y, Jurkowski TP, Stamerjohanns H, Reinhardt R, Jeltsch A (2008) Bisulfite sequencing Data Presentation and Compilation (BDPC) web server–a useful tool for DNA methylation analysis. Nucleic Acids Res 36(5):e34
90. Xi Y, Li W (2009) BSMAP: whole genome bisulfite sequence MAPping program. BMC Bioinformatics 10:232
91. Hackenberg M, Previti C, Luque-Escamilla PL, Carpena P, Martinez-Aroza J, Oliver JL (2006) CpGcluster: a distance-based algorithm for CpG-island detection. BMC Bioinformatics 7:446
92. Wang Y, Leung FC (2004) An evaluation of new criteria for CpG islands in the human genome as gene markers. Bioinformatics 20(7):1170–1177
93. Takai D, Jones PA (2003) The CpG island searcher: a new WWW resource. In Silico Biol 3 (3):235–240
94. Ioshikhes IP, Zhang MQ (2000) Large-scale human promoter mapping using CpG islands. Nat Genet 26(1):61–63
95. Ong SE, Mann M (2005) Mass spectrometry-based proteomics turns quantitative. Nat Chem Biol 1(5):252–262
96. Bantscheff M, Schirle M, Sweetman G, Rick J, Kuster B (2007) Quantitative mass spectrometry in proteomics: a critical review. Anal Bioanal Chem 389(4):1017–1031
97. Mann M, Ong SE, Gronborg M, Steen H, Jensen ON, Pandey A (2002) Analysis of protein phosphorylation using mass spectrometry: deciphering the phosphoproteome. Trends Biotechnol 20(6):261–268
98. Schlessinger J (2000) Cell signaling by receptor tyrosine kinases. Cell 103(2):211–225
99. Gygi SP, Rist B, Gerber SA, Turecek F, Gelb MH, Aebersold R (1999) Quantitative analysis of complex protein mixtures using isotope-coded affinity tags. Nat Biotechnol 17 (10):994–999
100. Thompson A, Schafer J, Kuhn K, Kienle S, Schwarz J, Schmidt G, Neumann T, Johnstone R, Mohammed AK, Hamon C (2003) Tandem mass tags: a novel quantification strategy for comparative analysis of complex protein mixtures by MS/MS. Anal Chem 75(8):1895–1904
101. Gafken PR, Lampe PD (2006) Methodologies for characterizing phosphoproteins by mass spectrometry. Cell Commun Adhes 13(5–6):249–262
102. Zhu W, Smith JW, Huang CM (2010) Mass spectrometry-based label-free quantitative proteomics. J Biomed Biotechnol 2010:840518
103. Ahrends R, Pieper S, Neumann B, Scheler C, Linscheid MW (2009) Metal-coded affinity tag labeling: a demonstration of analytical robustness and suitability for biological applications. Anal Chem 81(6):2176–2184
104. Munchbach M, Quadroni M, Miotto G, James P (2000) Quantitation and facilitated de novo sequencing of proteins by isotopic N-terminal labeling of peptides with a fragmentation-directing moiety. Anal Chem 72(17):4047–4057
105. Gruhler A, Schulze WX, Matthiesen R, Mann M, Jensen ON (2005) Stable isotope labeling of Arabidopsis thaliana cells and quantitative proteomics by mass spectrometry. Mol Cell Proteomics 4(11):1697–1709
106. Beausoleil SA, Jedrychowski M, Schwartz D, Elias JE, Villen J, Li J, Cohn MA, Cantley LC, Gygi SP (2004) Large-scale characterization of HeLa cell nuclear phosphoproteins. Proc Natl Acad Sci USA 101(33):12130–12135

107. Shadforth IP, Dunkley TP, Lilley KS, Bessant C (2005) i-Tracker: for quantitative proteomics using iTRAQ. BMC Genomics 6:145
108. Brill LM, Xiong W, Lee KB, Ficarro SB, Crain A, Xu Y, Terskikh A, Snyder EY, Ding S (2009) Phosphoproteomic analysis of human embryonic stem cells. Cell Stem Cell 5(2):204–213
109. Mortensen P, Gouw JW, Olsen JV, Ong SE, Rigbolt KT, Bunkenborg J, Cox J, Foster LJ, Heck AJ, Blagoev B et al (2010) MSQuant, an open source platform for mass spectrometry-based quantitative proteomics. J Proteome Res 9(1):393–403
110. Lin WT, Hung WN, Yian YH, Wu KP, Han CL, Chen YR, Chen YJ, Sung TY, Hsu WL (2006) Multi-Q: a fully automated tool for multiplexed protein quantitation. J Proteome Res 5 (9):2328–2338
111. Griffin TJ, Xie H, Bandhakavi S, Popko J, Mohan A, Carlis JV, Higgins L (2007) iTRAQ reagent-based quantitative proteomic analysis on a linear ion trap mass spectrometer. J Proteome Res 6(11):4200–4209
112. Perkins DN, Pappin DJ, Creasy DM, Cottrell JS (1999) Probability-based protein identification by searching sequence databases using mass spectrometry data. Electrophoresis 20 (18):3551–3567
113. Keller A, Eng J, Zhang N, Li XJ, Aebersold R (2005) A uniform proteomics MS/MS analysis platform utilizing open XML file formats. Mol Syst Biol 1:2005 0017
114. Chi SW, Zang JB, Mele A, Darnell RB (2009) Argonaute HITS-CLIP decodes microRNA-mRNA interaction maps. Nature 460(7254):479–486
115. Walhout AJ, Vidal M (2001) Protein interaction maps for model organisms. Nat Rev Mol Cell Biol 2(1):55–62
116. Hutchins JR, Toyoda Y, Hegemann B, Poser I, Heriche JK, Sykora MM, Augsburg M, Hudecz O, Buschhorn BA, Bulkescher J et al (2010) Systematic analysis of human protein complexes identifies chromosome segregation proteins. Science 328(5978):593–599
117. Pardo M, Lang B, Yu L, Prosser H, Bradley A, Babu MM, Choudhary J (2010) An expanded Oct4 interaction network: implications for stem cell biology, development, and disease. Cell Stem Cell 6(4):382–395
118. Crox (2007) ELAND: efficient large-scale alignment of nucleotide databases. Illumina, San Diego
119. Abelev BI, Aggarwal MM, Ahammed Z, Anderson BD, Arkhipkin D, Averichev GS, Bai Y, Balewski J, Barannikova O, Barnby LS et al (2008) System-size independence of directed flow measured at the BNL relativistic heavy-ion collider. Phys Rev Lett 101(25):252301
120. Li H, Ruan J, Durbin R (2008) Mapping short DNA sequencing reads and calling variants using mapping quality scores. Genome Res 18(11):1851–1858
121. Smith AD, Xuan Z, Zhang MQ (2008) Using quality scores and longer reads improves accuracy of Solexa read mapping. BMC Bioinformatics 9:128
122. Zhang Y, Liu T, Meyer CA, Eeckhoute J, Johnson DS, Bernstein BE, Nusbaum C, Myers RM, Brown M, Li W et al (2008) Model-based analysis of ChIP-Seq (MACS). Genome Biol 9(9):R137
123. Chen X, Xu H, Yuan P, Fang F, Huss M, Vega VB, Wong E, Orlov YL, Zhang W, Jiang J et al (2008) Integration of external signaling pathways with the core transcriptional network in embryonic stem cells. Cell 133(6):1106–1117
124. Eisen MB, Spellman PT, Brown PO, Botstein D (1998) Cluster analysis and display of genome-wide expression patterns. Proc Natl Acad Sci USA 95(25):14863–14868
125. Heintzman ND, Stuart RK, Hon G, Fu Y, Ching CW, Hawkins RD, Barrera LO, Van Calcar S, Qu C, Ching KA et al (2007) Distinct and predictive chromatin signatures of transcriptional promoters and enhancers in the human genome. Nat Genet 39(3):311–318
126. Hon G, Wang W, Ren B (2009) Discovery and annotation of functional chromatin signatures in the human genome. PLoS Comput Biol 5(11):e1000566

127. Spirin V, Mirny LA (2003) Protein complexes and functional modules in molecular networks. Proc Natl Acad Sci USA 100(21):12123–12128
128. Roguev A, Wiren M, Weissman JS, Krogan NJ (2007) High-throughput genetic interaction mapping in the fission yeast Schizosaccharomyces pombe. Nat Methods 4(10):861–866
129. Lu R, Markowetz F, Unwin RD, Leek JT, Airoldi EM, MacArthur BD, Lachmann A, Rozov R, Ma'ayan A, Boyer La et al (2009) Systems-level dynamic analyses of fate change in murine embryonic stem cells. Nature 462:358–362
130. Mitsui K, Tokuzawa Y, Itoh H, Segawa K, Murakami M, Takahashi K, Maruyama M, Maeda M, Yamanaka S (2003) The homeoprotein Nanog is required for maintenance of pluripotency in mouse epiblast and ES cells. Cell 113(5):631–642
131. Chambers I, Colby D, Robertson M, Nichols J, Lee S, Tweedie S, Smith A (2003) Functional expression cloning of Nanog, a pluripotency sustaining factor in embryonic stem cells. Cell 113(5):643–655
132. Mikkelsen TS, Hanna J, Zhang X, Ku M, Wernig M, Schorderet P, Bernstein BE, Jaenisch R, Lander ES, Meissner A (2008) Dissecting direct reprogramming through integrative genomic analysis. Nature 454(7200):49–55
133. Hawkins RD, Hon GC, Lee LK, Ngo Q, Lister R, Pelizzola M, Edsall LE, Kuan S, Luu Y, Klugman S et al (2010) Distinct epigenomic landscapes of pluripotent and lineage-committed human cells. Cell Stem Cell 6(5):479–491
134. Guttman M, Amit I, Garber M, French C, Lin MF, Feldser D, Huarte M, Zuk O, Carey BW, Cassady JP et al (2009) Chromatin signature reveals over a thousand highly conserved large non-coding RNAs in mammals. Nature 458(7235):223–227
135. Pandey RR, Mondal T, Mohammad F, Enroth S, Redrup L, Komorowski J, Nagano T, Mancini-Dinardo D, Kanduri C (2008) Kcnq1ot1 antisense noncoding RNA mediates lineage-specific transcriptional silencing through chromatin-level regulation. Mol Cell 32 (2):232–246
136. Nagano T, Mitchell JA, Sanz LA, Pauler FM, Ferguson-Smith AC, Feil R, Fraser P (2008) The Air noncoding RNA epigenetically silences transcription by targeting G9a to chromatin. Science 322(5908):1717–1720
137. Khalil AM, Guttman M, Huarte M, Garber M, Raj A, Rivea Morales D, Thomas K, Presser A, Bernstein BE, van Oudenaarden A et al (2009) Many human large intergenic noncoding RNAs associate with chromatin-modifying complexes and affect gene expression. Proc Natl Acad Sci USA 106(28):11667–11672
138. Kolasinska-Zwierz P, Down T, Latorre I, Liu T, Liu XS, Ahringer J (2009) Differential chromatin marking of introns and expressed exons by H3K36me3. Nat Genet 41(3):376–381
139. Andersson R, Enroth S, Rada-Iglesias A, Wadelius C, Komorowski J (2009) Nucleosomes are well positioned in exons and carry characteristic histone modifications. Genome Res 19 (10):1732–1741
140. Schwartz S, Meshorer E, Ast G (2009) Chromatin organization marks exon-intron structure. Nat Struct Mol Biol 16(9):990–995
141. Luco RF, Pan Q, Tominaga K, Blencowe BJ, Pereira-Smith OM, Misteli T (2010) Regulation of alternative splicing by histone modifications. Science 327(5968):996–1000
142. Spies N, Nielsen CB, Padgett RA, Burge CB (2009) Biased chromatin signatures around polyadenylation sites and exons. Mol Cell 36(2):245–254
143. Taylor J, Schenck I, Blankenberg D, Nekrutenko A (2007) Using galaxy to perform large-scale interactive data analyses. Curr Protoc Bioinform Chapter 10:Unit 10.15
144. Blankenberg D, Taylor J, Schenck I, He J, Zhang Y, Ghent M, Veeraraghavan N, Albert I, Miller W, Makova KD et al (2007) A framework for collaborative analysis of ENCODE data: making large-scale analyses biologist-friendly. Genome Res 17(6):960–964
145. Ashburner M, Ball CA, Blake JA, Botstein D, Butler H, Cherry JM, Davis AP, Dolinski K, Dwight SS, Eppig JT et al (2000) Gene ontology: tool for the unification of biology. The Gene Ontology Consortium. Nat Genet 25(1):25–29

146. Subramanian A, Tamayo P, Mootha VK, Mukherjee S, Ebert BL, Gillette MA, Paulovich A, Pomeroy SL, Golub TR, Lander ES et al (2005) Gene set enrichment analysis: a knowledge-based approach for interpreting genome-wide expression profiles. Proc Natl Acad Sci USA 102(43):15545–15550
147. Taslim C, Wu J, Yan P, Singer G, Parvin J, Huang T, Lin S, Huang K (2009) Comparative study on ChIP-seq data: normalization and binding pattern characterization. Bioinformatics 25(18):2334–2340
148. ENCODE Project Consortium (2011) A user's guide to the encyclopedia of DNA elements (ENCODE). PLoS Biol 9(4):e1001046

Chapter 8
High Throughput Affinity Purification and Mass Spectrometry to Determine Protein Complex Interactions

Anna Malovannaya, Rainer B. Lanz, Bert W. O'Malley, and Jun Qin

Abstract Affinity purification, or immunoprecipitation, followed by mass spectrometry (AP/MS or IP/MS, respectively) is a proper and powerful method to analyze protein complexes. Technological capabilities of mass spectrometry instrumentation have dramatically improved over the last decade, and now allow routine detection of high- and low-abundance components in composite protein mixtures, expanding our ability to use co-immunoprecipitation to study cellular regulatory protein interaction networks. In this review, we summarize several key efforts and accomplishments in applying mass spectrometry towards mapping of a proteome-wide interactome. Emerging high-throughput studies begin to etch a 'systems biology' view of protein-protein interactions, where modularity and interconnectivity of cellular protein machinery are unequivocally revealed. Yet, major challenges still remain in the application of high-throughput (HT) affinity purification for studies in higher organisms, particularly in human cells, and in addressing persistent issues associated with analysis, interpretation, and dissemination of massive IP/MS data. We describe some of the conceptual and practical challenges of the HT-IP/MS approach, and discuss feasible solutions in context of the protein complexes involved in transcriptional regulation.

Keywords Affinity purification • Immunoprecipitation • Mass spectrometry • Protein complexes • Protein-protein interactions

---

A. Malovannaya (✉) • R.B. Lanz • B.W. O'Malley • J. Qin
Baylor College of Medicine, One Baylor Plaza, BCM 130 Houston, TX 77030
e-mail: malovann@bcm.edu; rlanz@bcm.edu; berto@bcm.edu; jqin@bcm.edu

A. Ma'ayan and Ben D. MacArthur (eds.), *New Frontiers of Network Analysis in Systems Biology*, DOI 10.1007/978-94-007-4330-4_8,

8.1 Mass Spectrometry as a Discovery Tool for Identification of Protein Interactions: From Model Organisms to Higher Eukaryotes

Molecular mechanisms of cellular responses to external and internal stimuli are largely manifested by changes in the landscapes of protein-protein interactions. Over the last decade, proteomics mass spectrometry overcame many limitations in protein interaction discovery. The development of peptide-friendly ionization methods for mass spectrometry created a technology which allowed unbiased protein identifications to occur with unprecedented speed and sensitivity (for review see [66]). The identification of protein complexes that could be successfully purified benefited greatly from this technique.

Early success in biochemical isolation and identification of yeast and mammalian protein complexes by mass spectrometry, particularly those of RNA Polymerase-II Mediator complex [27, 40], SWI/SNF chromatin ATP-ases [71], and histone deacetylase complexes associated with HDAC1, -2, and -3 [22, 28], suggested that the majority of biological processes are performed by multi-molecular protein 'machines'. These discoveries fundamentally changed the ways in which we direct scientific investigations, ultimately encouraging the discovery-driven research we do today. This surge in numbers for mass spectrometry-based studies of protein interactions now promises to reveal ordered cellular interactomes that are conserved at the level of high-order protein interactions across seemingly disparate organisms.

Classically, a core protein complex is defined as a group of interacting proteins with stable stoichiometry over multiple column purifications (e.g. ion-exchange and Sepharose-based size exclusion). This type of strenuous biochemical procedure demands very high affinity between the protein subunits of the core complex module. While many stable complexes can endure co-fractionations, the majority of regulatory protein interactions are now known to be much weaker in nature. For example, protein interactions that are formed in response to activation of cellular pathways are generally inducible and also reversible. Two types of transient associations with obvious biological consequence are the enzyme-substrate interactions and the binding of two (or more) complexes to perform particular combinatorial or sequential functions. For the latter, initiation of transcription is a good example that involves extensive communication within the protein machinery, where DNA-binding transcription factors localize to particular genomic loci, subsequently recruiting chromatin remodeling complexes for the positioning of RNA Polymerase-II at the open promoter, and ultimately dissociating from the DNA to give way to RNA-processing enzymes and transcriptional terminators [74].

The composition of core modules defines the basic functional identity of a protein complex, while weaker higher order complex-complex interactions determine how and when these functions are utilized in different cellular contexts. Thus, a cell can then be viewed as an intricate network of meta-stable protein complex assemblies connected in a temporal series of dynamic associations [1, 36].

Affinity-based purification is a logical first choice for the isolation of interacting proteins. Its wide spread use has been hampered by inherent issues associated with non-specifically binding proteins, and the tendency of antibodies to cross-react with unintended 'secondary' antigens. Furthermore, large-scale AP/MS efforts generally have been limited to stringent purification procedures and tagging of bait proteins with a single type of affinity epitope, (e.g. Flag- or TAP-). Because epitope-tagging shows only one pattern of cross-reactivity, it previously has been considered a prerequisite for HT-AP/MS studies. The most comprehensive AP/MS studies of endogenous protein interaction networks were thus far done in yeast, where in-locus epitope-tagging of the complete ORFeome is feasible through homologous recombination [25, 37]. This type of genetic manipulation, however, is severely limited in human cell systems, possibly with the exception of a relatively new approach that approximates endogenous expression of tagged mammalian transgenes in a bacterial artificial chromosome (BAC) system [53]. Recent studies that focused on systematization of AP/MS for human protein complex identification, the largest published by Ewing and colleagues, were obtained with overexpression of epitope-tagged bait proteins [20]. Large-scale overexpression experiments are generally labor intensive and limited to moderate size proteins that are not toxic to the cell when overexpressed. When successfully expressed, recombinant protein levels rarely correspond to the endogenous levels of the native protein, which likely alter or induce non-native associations of questionable biological relevance. Furthermore, many regulatory proteins have the capacity to profoundly alter the cell state if their levels are changed. The complexity of the human proteome also is much higher than that of unicellular model organisms. The human reference assembly has more than 33,000 annotated genes, most of which produce alternative protein isoforms, and all of which are likely to be differentially expressed or have different functions across cell types, making inferences from model organisms and usage of methodologically intricate procedures particularly limited if not faulty. We invite the readers to review primary literature that addresses these issues by improving the efficiency of tagging procedures in mammalian cells and by regulating overexpression levels [8, 19, 23, 34, 53]. Clearly understanding human protein networks on the molecular level must involve the characterization of native – endogenous – protein complexes with immunoprecipitation by primary antibodies. This task needs an approach that overcomes the inherent complications of immunoprecipitation techniques to the extent that high-throughput and large-scale IP/MS studies become feasible. In this review, we will focus largely on key concepts and concerns in the analysis of endogenous protein interactions from HT-IP/MS studies. Furthermore, we highlight the prospects for using this approach in performing global comparative studies in any cell type of choice. These goals are particularly important for deciphering the biology of higher organisms.

8.2 Challenges in Purification of Regulatory Protein Complex Networks for Mass Spectrometry

8.2.1 Preservation of Transient Affinities in AP/MS: The 'Non-specific' Dilemma

Affinity purification can be performed at different stringencies, allowing purification of not only core complexes, but also their low affinity interactors, including regulatory proteins, which always co-precipitate at sub-stoichiometric levels. The sensitivity of current mass spectrometers is capable of supporting protein identification for mild isolations of 'extended' affinity purified complexes, as the detection of minor components and weaker complex interactors have become routine with newer systems. Under these isolation conditions, however, non-specific and cross-reacting proteins are invariably present and measurable by MS, which significantly impedes the interpretation of the data (Fig. 8.1). Generally, it is desirable to clear affinity isolations of non-specific binding, which up to now has been done mostly by increasing the stringency of purifications (e.g., through high salt or detergent concentrations in the washes) at a cost of losing weaker interactions. Clearly, as we expect to saturate the isolation (and hence identification) of stable core complex compositions in upcoming years, the future of AP/MS lies in the ability to systematically identify extended regulatory networks that do not have high enough affinity to sustain such high-stringency purifications. If the majority of specific transient affinities appear in similar ranges as the affinity of proteins that bind non-specifically, a major dilemma exists as to how non-specific binding can be reduced without loosing the weak interactions. As we will learn below, this is not solely accomplished by applying different experimental conditions, but is also solved *in silico* by identification of the non-specific interactions.

We have recently described the composition of abundant proteins in soluble and insoluble (precipitate, or aggregate) fractions of cellular extracts from HeLa S3 cells [44]. While monitoring the appearance of non-specific precipitation for establishing a robust IP protocol, we noticed that insoluble aggregation contributes most to so-called non-specific binding. In line with our observations, Cristea and colleagues showed that specific binding of the antigens does not improve significantly by prolonged incubation with antibody, but the amount of non-specific binders increases with incubation time [16]. Naturally, cellular lysates are metastable, and proteins will progressively aggregate when incubated at low temperature (4°C); aggregates easily adhere to beads and resist washing. These observations led to improvements in our immunoprecipitation protocol, where, for the first time, consistent reduction of non-specific components can be achieved by ultracentrifugation prior to binding of a protein immuno-complex to beads [44, 72]. Ultracentrifugation effectively pellets the majority of accumulated insoluble precipitates. Since this protocol can be performed in the absence of detergent or excess salt, it offers a particularly attractive solution for retention of weak associations in a protein complex.

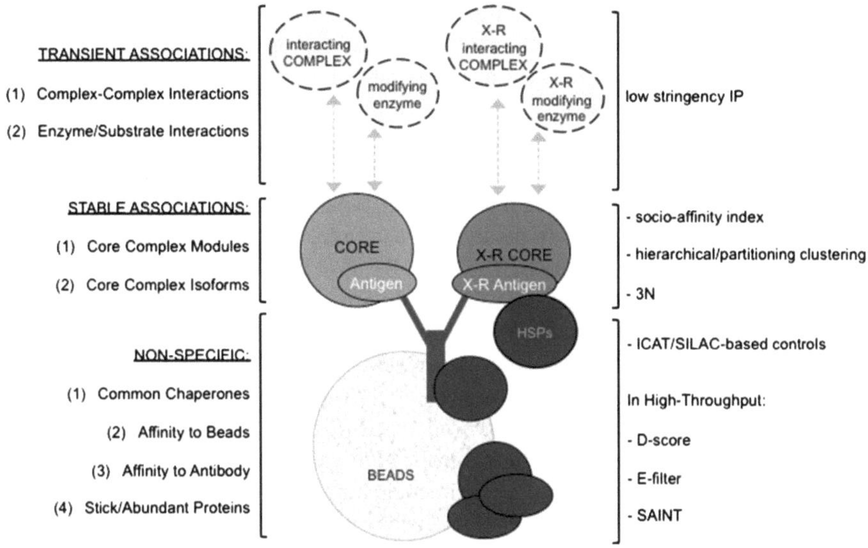

Fig. 8.1 Key challenges in analysis of protein complex interaction networks from IP/MS data. Affinity purification of protein complexes generally results in identification of several protein categories that represent different challenges for analyses of IP/MS results. First, proteins fall into specific and non-specific classes; where specific antibody affinities can be representing both targeted, as well as cross-reacting (X-R), antigens and their associations. Non-specific binding can be differentiated by experimental methods using isotope-labeling quantitative mass spectrometry, or, for large-scale high-throughput datasets, by statistical analysis. True protein interactions and composition of protein complexes are then inferred by subjecting data to standard or customized clustering analyses. Lastly, preserving transient interactions between two or more core complex modules is a major challenge for affinity-based protein interaction studies. Preservation of transient interactions can be routinely achieved by using a low stringency immunoprecipitation procedure with added ultracentrifugation step to separate non-specific precipitates. These methods are discussed in later sections of this chapter with references to primary literature

8.2.2 Experimental and Statistical Solutions for Determining Non-specific Binding in AP/MS

While an optimized immunoprecipitation protocol can significantly reduce non-specific proteins in the analytes, residual levels of non-specific proteins are likely to be retained by any procedure. This background non-specificity can be directly determined by quantitative mass spectrometry with isotope-labeled controls, or, alternatively, in large high-throughput datasets using statistical means.

Both stable isotope labeling in culture (SILAC) and isotope-coded affinity tag (ICAT) methods have been used to differentiate non-specific and cross-reacting binding proteins. For example, RNA polymerase II pre-initiation (PIC) complex has been measured with ICAT-labeled samples of PIC complexes assembled on promoter DNA in the presence or absence of TATA-binding protein (TBP) [54]. Similar logic is used in the I-DIRT method, where control precipitation is carried out in cells lacking the primary antigen and grown in differentially labeled SILAC media [68]. In both methods, heavy and light-isotope labeled peptides from non-specific proteins are detected at comparable levels, while peptide abundance of true interacting proteins is significantly enriched in comparison to control. An interesting extrapolation of such methods is the inclusion of titrated isotope-labeled samples, which alleviates some of the issues associated with dynamic ranges of peptide quantification, and providing better accuracy than the straightforward repetition of MS measurements. For example, SILAC-based dilutions were used with success to study the dynamics of the Foxo3A interactome in full serum conditions and upon cell starvation and inhibition of PI3K pathways [56].

While these methods are useful for definitive investigation of interaction specificity, scaling up quantitative controls to match the volume of high-throughout AP/MS projects is time- and cost-prohibitive. Fortunately, when large datasets are available, statistical analyses can be used to substantially reduce the necessity for extensive traditional experimental controls. Non-specific proteins are expected to appear across multiple unrelated experiments, and often are eliminated based on a straightforward frequency cutoff. The frequency parameters also are used indirectly in analyses that assess likelihood of protein-protein associations in HT-AP/MS datasets. For example, a PPI socio-affinity index considers frequencies of protein identifications in a given dataset to determine chances that any two proteins will co-precipitate a certain number of times [25]. Proteins that co-precipitate more than expected by chance are logically assumed to associate with each other in a biologically relevant interaction. Socio-affinity indices also automatically penalize frequently occurring proteins, allowing use of raw datasets for clustering analysis. However, penalizing frequent proteins in isolates can be counterintuitive in biased datasets where particular proteins are overrepresented by design (for example, when they are the intended targets in IPs).

There is now an emerging trend to use spectral counts (SPCs) of peptide hits and their derivatives, such as normalized spectral abundance factor (NSAF), as semi-quantitative measurements of protein recovery in MS [48, 49, 51]. Filtering data for likely specific identifications then becomes a task of defining the background at which each given protein appears in a large number of biologically unrelated experiments [10, 44, 65]. Overall, per protein non-specificity thresholds supersede earlier frequency-based solutions. In studies that utilize epitope-tagging, the majority of data cleaning is concerned with elimination of non-specific proteins. For HT-IP/MS studies, however, a larger problem is 'specific' cross-reactivity of primary antibodies, as discussed in the next section.

8.2.3 Overcoming Variable Cross-Reactivity of Primary Antibodies

Affinity immunoprecipitation purification of protein complexes with primary antibodies to endogenous proteins allows the investigation of native protein interactions in higher eukaryotes. Immunoprecipitation using primary antibodies, however, was not considered for high-throughput AP/MS until recently, because rigorous characterization and availability of strictly specific antibodies for particular protein targets were considered a prerequisite for the experiment. In the absence of corresponding controls for all antibodies, such data were expected to be difficult to evaluate. However, such expectation was proven to be incorrect. With persuasion that protein interaction proteomics will inadvertently have to confront these issues, we acquired a large dataset of IP/MS experiments for transcriptional regulators and proposed an analysis approach that effectively separates cross-reacting proteins in IP/MS results [44]. In fact, the solution comes in a form of relatively simple reciprocity constraint. Reciprocal immunoprecipitation, generally done with paired IPs followed by Western blot of interacting proteins, is a commonly used 'gold standard' test for verification of interactions in hypothesis-driven research. The concept of the interaction reciprocity in a given dataset also can be generalized to describe the number of antibody-different experiments where two given proteins are co-found. The non-reciprocal associations that co-precipitate with only one antibody contain the vast majority of likely cross-reacting proteins and can be omitted or flagged during analyses, albeit at cost of many weak true interactors. Thus, increasing reciprocity would generally indicate more confidence in an association between proteins, given that non-specific identifications are filtered out by means described above. Furthermore, it is not necessary that the reciprocal analyses of a particular protein association be directed towards subunits of the same protein complex. These antibodies can be targeted to the same antigen, another component of the protein complex, an interacting complex, or even unrelated proteins, as long as the protein association is observed. Since the problem of cross-reactivity lies in the assumption that all identified proteins are associated with the antigen (or 'bait') to which the antibody was raised, this problem becomes less of an issue if we do not rely strictly on 'bait/prey' relationships in calling protein complexes. When larger datasets of HT-IP/MS data become available, the reciprocity thresholds can be adjusted to account for higher chance of encountering reciprocal cross-reactivity.

Finally, since cross-reactivity is just repeated in same-antibody experiments, data acquisition in a HT project should be based on diversification of antibody affinities rather than on repetition of the same-antibody experiments, which also benefits workload without significant loss of information.

8.3 Deriving Protein Complexome from AP/MS Data

The complexity of AP/MS data ventures far beyond cleansing data of the non-specific and cross-reacting identifications. In comparison to two-hybrid and bimolecular fluorescence complementation (BiFC) methods, where the outcome is largely Boolean in nature (interactions are either true or false), affinity purification often preserves the hierarchy of protein binding. As such, stable interactions are biochemically distinct from transient interactions in affinity isolations, and frequent interactions are more prominent than rare events. This information about organization of PPI networks is contained in the relative abundances of co-purified proteins. Here again, the spectral counts prove to be a useful indicator of relative protein levels. The estimation of relative levels at which two proteins or two complexes co-purify helps the classification of interactions into stable and transient – or frequent and rare – categories. One has to consider, however, at least two confounding limitations in evaluation of such information: (1) protein complexes will partially dissociate during the isolation, and (2) there is a possibility of incidental antibody-mediated displacement, even for epitope-tagged baits. It is likely that these technical complications could be overcome by deep reciprocal coverage. Ultimately, however, the availability of very large AP/MS studies that are internally consistent will determine the extent of informational content of affinity-based datasets.

The most basic task in deconvolution of HT-IP/MS data is the allocation of proteins into modules that represent stable core protein complexes. Protein complexes can be defined here as a group of proteins in a set of affinity isolations that display similar behavior in a given clustering approach. Since core protein complexes are easiest to preserve by affinity purification, their compositions also are the easiest, comparatively speaking, to decipher. As expected, multiple computational approaches that rely on repeated co-purification of core complex subunits and assessment of relative levels of proteins in the immuno-complexes have been shown to be effective for analyses of core complex composition. These include, but are not limited to, both hierarchical clustering and partitioning (e.g. K-means, fuzzy clustering) algorithms, which can utilize a variety of standard distance metrics (e.g. Pearson, Euclidian, Manhattan). There are ample choices of reviews and primary literature concerned with application, evaluation, and comparison of these methods [15, 25, 32, 36, 62]. Individual approaches may have both strengths and weaknesses that are dependent on the type of knowledge and task at hand. Effectiveness of statistical models, for example, relies on the extent and evenness of protein coverage, and partitioning requires an expectation of cluster – or protein complex – numbers, which is rarely certain. To this end, many studies will have to resort to iterative variation of clustering methods and parameters [62]. A composite socio-affinity index score also has been defined for AP/MS data and used successfully in conjunction with hierarchical clustering to deconvolute the yeast proteomic complexome [25]. Interestingly, socio-affinity indices effectively maximize information gain from raw AP/MS data by combining separate metrics that gauge protein associations against complementary 'spoke' and 'matrix' interactive models [4].

The majority of analyses performed to date rely on bait-prey data arrangement, knowledge of which is often not explicitly possible for IP/MS data. We recently described a logical framework for Near-Neighbor Network (3N) analysis that relies on sequential constraining of a data pool to define a protein complex independent of bait-prey relationships [44]. This analysis is a particularly attractive option for HT-IP/MS studies because it avoids the necessity to characterize, or for that matter, to know the actual antigens of primary antibodies. In general, 3N can be thought of as an application of 'matrix' interaction logic, which builds upon characteristic biochemical behavior of protein complexes to narrow down groups of proteins belonging to the core modules. Below we briefly discuss major considerations in application of 3N analysis for data mining of HT-IP/MS datasets.

8.3.1 Near-Neighbor Network (3N) Analysis Schema for Antigen-Independent Protein Complex Assignment in HT-IP/MS

For the derivation of core protein complex compositions, the underlining premise is that true complex components should co-occur in multiple reciprocal IPs, especially when at least one of the subunits is identified abundantly by mass spectrometry. In fact, this is a basic concept of stoichiometric interdependence, where the opposite is also true: when one of the components of the core protein complex is identified at a low level, it is likely that the other components are going to have low abundance or, more importantly, may even fall below the detection limit of mass spectrometry. This implies that interpretation of protein associations in HT-IP/MS data needs to be aware of technical limitations, particularly, in imposing penalties on identifications that are likely missed. The 3N approach rests on this basic definition of a protein complex as a stoichiometrically interdependent assembly – the analysis is repeated for each protein individually, and uses several essential constraints discussed next. First, the deconvolution of large IP/MS datasets can be improved and simplified by iterative sampling of the proteins' 'top experiment' subsets (where a given protein of interest is present at the highest levels in the data). This minimizes the influence of experimental variations, especially in cases where complex subunits fall below detection threshold. Second, to solve cross-reactivity concerns in HT-IP/MS, reciprocal co occurrence is emphasized by requiring co-presence in multiple antibody-different IPs, not just in experimental repeats. Lastly, core complex subunits are differentiated from other types of interactions based on a correlation of their SPCs, which asserts the degree of stoichiometric interdependency. When compared across multiple experiments, simple measures of similarity are sufficient to extract discernable correlation trends among proteins. For example, cosine similarity measures an angle between the SPC vectors, where protein SPC values per experiment are represented as vector coordinates in a multidimensional IP space. We have used this approach effectively to assess relative stoichiometry in isolated steady-state complexes of transcriptional coregulators, returning multiple known interactions with high accuracy [44]. Because there are no assumptions of bait-prey relationships in this approach, instances of protein complex identifications that are not targeted (such as cross-reactivity and complex-complex interactions) can be used for analysis, since they, too, exhibit preservation of intra-complex stoichiometry. In this way, cross-reactivity provides 'bonus' protein complex identifications and can be advantageous in a quest for proteome-wide coverage of endogenous interaction maps.

8.3.2 SPC-Based 3N Analysis Imposes Sufficient Constrains on Spatial Arrangement of Mediator Subunits

In addition to protein composition, it is often desirable to know the spatial arrangement of subunits within a protein complex. Interestingly, affinity purification data may be sufficient for hinting at the three-dimensional structure of protein complexes. Bioinformatics approaches have been proposed for the extraction of subunit arrangement within a protein complex from AP/MS data in combination with, or without, Y2H maps [24, 63]. 3N-based ranking is also surprisingly accurate at predicting interactions that are likely to be proximal. For example, the transcriptional regulator Mediator complex is used here to illustrate that extensive IP/MS data can have sufficient informational content to predict known arrangement of the subunits.

Both electron microscopy and Y2H maps have been used to define the structural arrangement of subunits in the Mediator complex [12, 18, 29, 67]. This large complex has at least four prominent modules: kinase, head, middle and tail [7, 18]. The closest neighbors of MED12 in 3N analysis (using 10–20 MED12-contating experiments) are MED13 and CCNC, which, in fact, are all the known components of the CDK8/8L kinase sub-module of the Mediator (Fig. 8.2a, adapted from Guglielmi et al. and Bourbon; and Fig. 8.2b). 3N-based ranking of Mediator subunits from three remaining modules – Head, Middle, and Tail – also can mimic literature-based arrangement of these sub-complexes (Fig. 8.2c–e). For the 'head' component MED18, MED27, -20, -6, -8, -4, -17, -30 and -14 are in closest 3N proximity. The top six of these MED18 neighbors are indeed in agreement with the model proposed (Fig. 8.2c). The same holds true for the MED21 subunit of the 'middle' module and 'tail' subunit MED16, which shows that it is proximal to MED24, -17, -15, -14, -25, -29, -1, -4 and POLII (Fig. 8.2d). In one contrast to the current model for the active complex, our analysis put MED16 of the Tail module and MED17 of the Head module in close proximity. This can mean that we recover a Mediator complex in its closed conformation, a scenario that is likely to happen when this coregulator is stripped from chromatin during nuclear extraction.

While determining protein complex organization from HT-IP/MS data is a largely uncharted area (mostly due to relative scarceness of HT-IP/MS data), it is interesting to speculate about the origin of these unexpected insights. In a sample population of an isolated multi-protein complex, the observed stoichiometry is perturbed differently between proximal and distal subunits. The extent to which two proteins dissociate during purification depends on the strength of interactions and the number of interaction surfaces between the two subunits (direct vs. indirect). In a simple model of a homogeneous core complex module, the direct binder can only be lost by disturbing one or few primary PPI interfaces, while an indirect binder can dissociate upon loss of any of a number of interactions needed for its preservation. Thus, on average, dissociation of indirect binders should happen more frequently than dissociation of direct binders. In terms of subunit abundances, or what we call the observed stoichiometric ratio, indirectly binding subunits should have a lesser SPC correlation to a protein of choice than its direct binders.

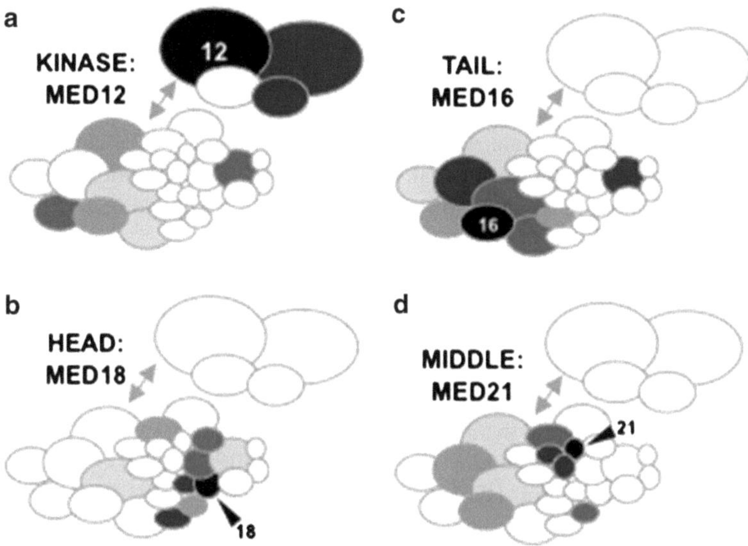

Fig. 8.2 3N-based ranking of near neighbors suggests spatial organization of Mediator subunits that is in agreement with literature models. Near neighbor assignments from 3N analyses of HT-IP/MS dataset (from data current to [44]) for (**a**). Kinase subunit MED12; (**b**). Head subunit MED18; (**c**). Tail subunit MED16; and (**d**). Middle subunit MED21 of the Mediator complex. The seed proteins are in *black*; *top* eight nearest subunits to each seed are color-coded based on their proximity to the seed protein as assigned by the 3N analysis, *darker shades* represent closer subunits. The *shading* is superimposed on the model proposed in literature [7] and reflects high degree of agreement

Thus, the similarity between protein SPC vectors can be used as a parameter for arranging 'neighbors' in the core complex. Better correlations between SPCs or abundance factors of two different proteins in a series of immunoprecipitations will favor directly interacting complex components.

8.3.3 *Modularity and Interconnectivity of the Cellular Complexome*

8.3.3.1 Core Complex Heterogeneity

It has become a common theme that a protein can reside in different protein complex 'isoforms', where a common set of 'core' subunits is coupled to different 'attachment' modules. Widespread protein complex heterogeneity has been described in yeast, where iterative clustering with different stringency of parameters revealed ~500 complexes with up to ~5,500 possible protein complex variants [25]. While it is likely that the majority of mathematically defined complex variants are not completely independent of each other, or for that matter, exactly accurate in predicting composition of individual biological complexes, this seminal study accented the importance of protein complex heterogeneity in biology and invited development of analytical methods that emphasize differentiation of such largely similar protein complexes. Multiple hypothesis-driven and several detailed medium-scale AP/MS studies in mammalian cells led to a conclusion that diversification of biological function through combinations of different core modules into higher-order protein complexes is a fundamental biological phenomenon [13, 42, 61].

For example, histone deacetylases HDAC1 and HDAC2 reside in several distinct corepressor complexes. Three major HDAC1/2 complexes have been described: nucleosome remodeling and deacetylase (NURD) complex, KDM1-containing (CoREST/BHC) histone demethylase complexes, and the SIN3A-associated (SAP) complex. We have reported deconvolution of these corepressor complexes from HT-IP/MS data using the 3N approach [44].

The NURD complex contains chromodomain helicases CHD3/CHD4, MTA family proteins, and core HDAC1/2 interactors RBBP7/RBBP4 and MBD2/MBD3. Furthermore, the proteins comprising this complex display differential expression in different tissues, suggesting a high degree of heterogeneity in subunit composition. Direct evidence also suggests that MBD2 and MBD3-containing NURDs are exclusive of each other [9, 39]. The NURD complex interacts with the SWI/SNF ATPase P-BAF complex, combining several activities that affect transcriptional processes in a 'super'-complex with the capacity for large-scale chromatin remodeling [21, 44, 76]. The NURD and CoREST complexes have been reported to functionally interact [73], but the CoREST module can be purified with HDAC1/2 independently of NURD complex subunits. In contrast to NURD and SAP complexes that have a very similar pool of proteins in different IPs, the majority of proteins that frequently associate with the CoREST module do not co-appear in different isolations, indicating that diverse sets of CoREST 'isocomplexes' with different attachments are more likely to exist than one large CoREST complex.

With the heterogeneity and modularity of the cellular interactome being a rule rather than an exception, there are ample examples in the primary literature that illustrate how basic functions of a 'core' module can be adapted towards different cellular tasks by attachment of auxiliary protein complex modules. For example, the eukaryotic clamp loader complex (RFC2/3/4/5) can utilize at least four different attachments – RFC1, RAD17, CHTF18, or ATAD5. These attachments dictate preferences of the RFC core for DNA substrates (e.g. double or single-stranded) and alternative DNA clamps (e.g. PCNA and 9-1-1 complexes); and these four RFC 'isocomplexes' are differentially utilized in processes associated with DNA synthesis in S-phase, DNA damage response, and maintenance of chromosome cohesion (for review see [6]). While characterization of RFC complexes required years of work by different laboratories; more recent medium-scale AP/MS studies have undertaken systematic isolation and functional characterization of heterogeneity in 'local' networks of core protein complexes.

The compositions of two core protein complexes that contain MYST1 histone (H4) acetyltransferase – male-specific lethal (MSL) and non-specific lethal (NSL) – have been determined to certainty by purification of nine different protein baits representing different subunits of these two complexes [13]. In the same study, the authors were able to show that MYST1 specificity is context-dependent of MSL and NSL. These examples highlight the unequivocal necessity to accurately split protein interaction networks into exclusive protein complexes that, in turn, hold the key to a better understanding of diversity and specificity of protein functions.

Protein complex heterogeneity, however, poses certain challenges for network analysis. The ability to differentiate similar (though separate) complexes just from affinity purification data is highly dependent on the density of experiments for the proteins that are common between different isocomplexes. Experiments that target shared subunits should recover multiple protein complexes, and a high density of such experiments ultimately favors merging of complex variants. Also, optimization of clustering parameters is needed to be able to differentiate these complexes as separate entities. Sardiu and colleagues acquired AP/MS data for RUVBL1/L2-containing complexes and showed that statistical separation of protein complex attachments is counter-balanced by the breakage of core complexes into smaller modules [61, 62]. As the complexome is catalogued, we will likely resort to clustering as a first step in annotation of core complexes, where imperfections of computationally derived predictions can be reconciled by manual inspection of local protein complex data and incorporation of knowledge from targeted functional studies.

8.3.3.2 Transient Protein Complex Interactions

The last conceptual tier of physical interactions is complex-complex interactions (CCIs). CCIs constitute what we could call the 'regulatory' complexome of the cell, or a set of interactions that are responsible for the dynamics of cellular responses to biological signals. These interactions are generally of a transient nature, which differentiates them functionally and biochemically from stable protein core complexes.

For enzyme/substrate interactions, there is a useful affinity and mass spectrometry-based approach that targets the substrates directly, rather than the interactions. For example, over 900 putative phospho sites on ~700 putative substrates of the ataxia-telangiectasia-mutated (ATM) and ATM-related (ATR) DNA damage sensor kinases have been identified using multiple antibodies raised to the known sites of ATM and ATR phosphorylation [45, 47]. In this experimental setup, the logical assumptions are that (1) an enzymatic pocket of a particular kinase has preference for a particular physical structure of the modification site embedded in the consensus amino acid 'motif', and (2) a phospho-antibody that mimics a kinase should be able to bind multiple substrates. Identification of kinase substrates is particularly amenable to this method; however, it remains to be seen whether substrates of enzymes for other types of post-translation modifications, such as acetylation and methylation, can be effectively isolated by this approach.

For identification of interactions between protein complex modules that are not of an enzyme/substrate nature, but come together to perform certain cellular tasks – such as the NURD/P-BAF or NURD/CoREST complex networks that remodel chromatin – reciprocal IP/MS of multiple protein complex subunits is one of the best experimental strategies. However, sub-stoichiometric recovery of 'secondary' interacting protein complexes through a primary antigen-containing complex presents additional challenge for bioinformatics analyses. When a protein complex is retrieved at low abundance, it is likely that a portion of the subunits of a complex will fall below the detection limit of mass spectrometry – a problem that surfaced in our discussion of core complex assignment from experiments with subpar recovery. In this case, bioinformatics that utilize binary protein-protein association logic are likely to penalize sub-stoichiometric CCIs, and additional analyses that model and evaluate the likelihood of interactions between partially represented complex modules may be needed.

8.4 HT-IP/MS in the Omics Era: The Final Frontier

As we are in the 'omics' phase of biomedical research that routinely pushes the limits of genome- and proteome-wide coverage, devising meaningful approaches that reconcile these diverse types of data becomes ever more pressing [3, 58, 75]. Protein complex interaction mapping stands to provide invaluable information for interpretation of expression profiling, ChIP-seq data, and genomic analysis of variations in SNPs and gene copy numbers. These studies generally rely on subsequent analyses of disease or pathway associations, whereas protein interaction data are largely self-defining. Next, we highlight a few imminent directions for integration of high-content studies, with a particular focus on the roles of HT-IP/MS in these endeavors.

8.4.1 Saturating the Interactome

With scalable methodology for identification of epitope-based and endogenous protein complexes, it is now feasible to acquire dense IP/MS data for most human proteins and to decipher their interaction networks in great detail. Still, binary yeast- and mammalian-two-hybrid (Y2H, M2H) and bimolecular fluorescence complementation (BiFC) screens can complement HT-IP/MS approaches in ordering protein interactions within protein complexes and in covering transient interactions missed during purification. Both two-hybrid and BiFC methods have the potential to cover the majority of direct binary protein interactions in the cells [46, 55, 70]. In turn, we can rely on affinity purifications to pose these

PPIs within the modular context of the complexome, where many more proteins associate indirectly.

Knowledge of direct binding partners can propel our ability to predict structural arrangement of multi-subunit protein complexes. For example, we used the well-studied Mediator complex to illustrate an enticing possibility that HT-IP/MS analysis such as 3N can, in fact, constrain spatial arrangement of protein complex subunits. These approximate predictions of proximal subunits, together with direct PPIs determined by binary protein interaction studies, could be sufficient for modeling high-order organization of protein complexes with limited structural information.

Interaction screens that use over-expression could have advantage for mapping transient interactions, in which exogenous expression of two interacting proteins at similarly high levels will enhance the detection of their interactions. Despite our best efforts to preserve weak interactions in affinity purification, routine identification of transient interactions in AP/MS is difficult and likely captures the most frequent interactors only. Han and colleagues, while working with a Y2H interaction data, proposed to differentiate proteins that interact with a large number of different molecules – 'date' hubs – from 'party' hubs that have a preferential and often restricted pool of interacting partners [31]. In AP/MS, 'party' hubs are analogous to the stable and well-preserved multi-subunit complexes. The most challenging proteins for AP/MS are those that display 'promiscuous' binding patterns of 'date' hubs. A biological interpretation of this phenomenon is that 'date' hubs stand at the crossroads of multiple molecular pathways, which may happen at different times, in different cellular contexts, and in a signal-dependent manner. It is reasonable to expect that 'date'-type proteins will suffer in an AP/MS approach, because in a population of cycling cells, each 'dating' partner represents only a minor fraction of the 'date's' total steady-state interaction network. However, targeted IP/MS studies that investigate these types of proteins in particular cell types, tissues, or under proper induced conditions is a lucrative and likely fruitful direction for studying the 'date' hubs. Furthermore, knowing that a given protein is of 'date' type might be used to justify a lower stringency of data analyses. For example, 'dating' interactions are very likely to be less repeatable and always sub-stoichiometric in AP/MS. Consequently, close (and often manual) interrogation of raw data can be utilized to retrieve potential interactors of 'date' hubs from HT-AP/MS data.

8.4.2 Integrating Protein Complexome Knowledge with Genome-Wide Sequencing Studies

At least two orthogonal 'omics' approaches can benefit from the availability of a proteome-wide protein complex map – chromatin immunoprecipitation followed by deep sequencing (ChIP-Seq) and genome-wide association studies (GWAS).

Chromatin immunoprecipitation followed by sequencing is a powerful technique to assay the interactions between proteins and DNA. The loci and dynamics of binding to chromatin for proteins that regulate gene transcription or maintain DNA and chromatin organization can be determined on a genome-wide scale [52]. Overlapping ChIP-Seq profiles of different proteins can lead to a better understanding of the combinatorial patterns for gene regulation. For example, studies of DNA-binding of the transcription factors FOXA1, estrogen (ERα) and androgen (AR) nuclear receptors, and the transcriptional coregulator KDM1 in MCF7 and LNCaP cells suggested that cell-specific patterns of gene expression are defined by combinatorial recruitment of these proteins to cell-specific chromatin loci [41]. Furthermore, we have outlined an approach for integration of information from ChIP-Seq, interaction proteomics, and mRNA expression profiling that allowed us to discover a new transcriptional network for the steroid receptor coactivator 3 (SRC-3/AIB1/pCIP/NCOA3) that involved RFX1 in activation of ER and SRC3-responsive genes [38]. An interesting direction for integration of proteomics and ChIP-based approaches would be to test recruitment of whole transcriptional protein complexes to the sites of gene regulation. Such studies are particularly beneficial for determining cellular functions for proteins that reside in multiple core complexes. Here, proteomics can be used to determine a set of protein complex subunits to be targeted by ChIP that are likely to yield discriminative information. One can then imagine that chromatin recruitment patterns for common 'core' and variable 'attachment' subunits can serve as a powerful indicator of functions for separate protein 'isocomplexes'. The emergence of studies that utilize interaction proteomics in conjunction with ChIP-sequencing also highlights the importance of developing HT-IP/MS reagents and techniques. It is likely that the most insightful research will require comparisons of multiple different cell types or tissues, where epitope-tagging and overexpression of exogenous baits adds labor and poses additional, and perhaps cell type-specific, challenges.

Another area of biomedical research where knowledge of protein interactions can aid interpretation of the data is genome-wide association studies (GWAS). GWAS are used to characterize prevalence of particular genetic variations in human populations in regards to diseases [5, 26, 60, 64, 69]. These studies are highly effective in finding significant correlations between a mutation and a trait, but have lesser predictive power for mechanisms of action and functional consequences of discovered genetic alterations. For example, the occurrence of particular somatic gene mutations in a tumor could be a consequence ('passenger'), rather than the cause ('driver'), of the disease [26, 30, 64]. Since etiologies of malignancies involve perturbations of specific signaling and transcriptional cascades, mutations in genes that code for protein constituents of these pathways are more likely to be causative. While the potency of IP/MS and PPI data in narrowing down human 'cancer gene' candidates has yet to be fully tested, grouping of functions and phenotypic consequences within protein complex boundaries was observed in yeast where deletion of different subunits of the same complex often results in similar phenotypic patterns [25]. Re-analyses of data from this latter study and from another yeast complexome dataset also suggested that essential gene products tend to group in protein complexes rather than distribute randomly among them, strongly favoring a unitary functional behavior for core protein complex modules [32, 37]. Thus, interpretation of the significance of analyses of GWAS data that aim to differentiate 'driver' and 'passenger' mutations can be greatly complemented by understanding the precise organization of genomic protein associations within protein complexes and their interaction networks.

8.5 From PPIs to CCIs: Implications for Protein Interaction Resources

In summary, a map of protein-protein interactions is now evolving into a modular network of protein complex interactions. Multiple lines of evidence suggest that protein complexes, rather than individual proteins, serve as units of biological function in the cell. To this end, major experimental efforts are now underway to establish and refine a consensus database of complex-complex interactions [43]. While it is likely that the implementation of mass spectrometry in high-throughput protein interaction studies will remain limited to a small number of dedicated centers and laboratories with extensive expertise in this technology, the information uncovered in these studies will be of great value to a majority of laboratories. Accordingly, there are multiple protein interaction databases, including BIND [33], BioGRID [11], DIP [59], MINT [14], HPRD [35], IntAct [2], CORUM [57], and MIPS [50], most of which now also integrate the PPIs discovered by mass spectrometry proteomics (for review also see [17]). However, representation of the interactome as a set of binary interaction PPI pairs still is limiting in conveying organization and preferences within higher-order complex-complex interactions that can be derived from HT-AP/MS. The interpretation of the proteomic interactome as a network of CCIs could significantly simplify our quest to understand molecular mechanisms of cellular life. Consequently, the development of dedicated resources for gathering and disseminating such information is a critical task for future protein interaction research.

Acknowledgements We thank Dr. Yaroslava Bulynko for critical reading of the manuscript. This work was supported by the NIH grants U19-DK62434 (including the Proteomics Strand funding to B.W.O. and J.Q. and the Collaborative Bridging Project funding to R.B.L.), NIDDK DK59820 (to B.W.O.), and NCI CA84199 (to J.Q.). We also acknowledge the support of the McLean Foundation through the Center for Molecular Discovery at Baylor College of Medicine.

References

1. Alberts B (1998) The cell as a collection of protein machines: preparing the next generation of molecular biologists. Cell 92:291–294
2. Aranda B, Achuthan P, Alam-Faruque Y, Armean I, Bridge A, Derow C, Feuermann M, Ghanbarian AT, Kerrien S, Khadake J et al (2010) The IntAct molecular interaction database in 2010. Nucleic Acids Res 38:D525–D531

3. Auffray C, Chen Z, Hood L (2009) Systems medicine: the future of medical genomics and healthcare. Genome Med 1:2
4. Bader GD, Hogue CW (2002) Analyzing yeast protein-protein interaction data obtained from different sources. Nat Biotechnol 20:991–997
5. Beroukhim R, Mermel CH, Porter D, Wei G, Raychaudhuri S, Donovan J, Barretina J, Boehm JS, Dobson J, Urashima M et al (2010) The landscape of somatic copy-number alteration across human cancers. Nature 463:899–905
6. Bloom LB (2009) Loading clamps for DNA replication and repair. DNA Repair (Amst) 8:570–578
7. Bourbon HM (2008) Comparative genomics supports a deep evolutionary origin for the large, four-module transcriptional mediator complex. Nucleic Acids Res 36:3993–4008
8. Bouwmeester T, Bauch A, Ruffner H, Angrand PO, Bergamini G, Croughton K, Cruciat C, Eberhard D, Gagneur J, Ghidelli S et al (2004) A physical and functional map of the human TNF-alpha/NF-kappa B signal transduction pathway. Nat Cell Biol 6:97–105
9. Bowen NJ, Fujita N, Kajita M, Wade PA (2004) Mi-2/NuRD: multiple complexes for many purposes. Biochim Biophys Acta 1677:52–57
10. Breitkreutz A, Choi H, Sharom JR, Boucher L, Neduva V, Larsen B, Lin ZY, Breitkreutz BJ, Stark C, Liu G et al (2010) A global protein kinase and phosphatase interaction network in yeast. Science 328:1043–1046
11. Breitkreutz BJ, Stark C, Reguly T, Boucher L, Breitkreutz A, Livstone M, Oughtred R, Lackner DH, Bahler J, Wood V et al (2008) The BioGRID interaction database: 2008 update. Nucleic Acids Res 36:D637–D640
12. Cai G, Imasaki T, Takagi Y, Asturias FJ (2009) Mediator structural conservation and implications for the regulation mechanism. Structure 17:559–567
13. Cai Y, Jin J, Swanson SK, Cole MD, Choi SH, Florens L, Washburn MP, Conaway JW, Conaway RC (2010) Subunit composition and substrate specificity of a MOF-containing histone acetyltransferase distinct from the male-specific lethal (MSL) complex. J Biol Chem 285:4268–4272
14. Ceol A, Chatr Aryamontri A, Licata L, Peluso D, Briganti L, Perfetto L, Castagnoli L, Cesareni G (2010) MINT, the molecular interaction database: 2009 update. Nucleic Acids Res 38: D532–D539
15. Choi H, Kim S, Gingras AC, Nesvizhskii AI (2010) Analysis of protein complexes through model-based biclustering of label-free quantitative AP-MS data. Mol Syst Biol 6:385
16. Cristea IM, Williams R, Chait BT, Rout MP (2005) Fluorescent proteins as proteomic probes. Mol Cell Proteomics 4:1933–1941
17. De Las Rivas J, Fontanillo C (2010) Protein-protein interactions essentials: key concepts to building and analyzing interactome networks. PLoS Comput Biol 6:e1000807
18. Dotson MR, Yuan CX, Roeder RG, Myers LC, Gustafsson CM, Jiang YW, Li Y, Kornberg RD, Asturias FJ (2000) Structural organization of yeast and mammalian mediator complexes. Proc Natl Acad Sci USA 97:14307–14310
19. Drakas R, Prisco M, Baserga R (2005) A modified tandem affinity purification tag technique for the purification of protein complexes in mammalian cells. Proteomics 5:132–137
20. Ewing RM, Chu P, Elisma F, Li H, Taylor P, Climie S, McBroom-Cerajewski L, Robinson MD, O'Connor L, Li M et al (2007) Large-scale mapping of human protein-protein interactions by mass spectrometry. Mol Syst Biol 3:89
21. Feng Q, Zhang Y (2003) The NuRD complex: linking histone modification to nucleosome remodeling. Curr Top Microbiol Immunol 274:269–290
22. Fleischer TC, Yun UJ, Ayer DE (2003) Identification and characterization of three new components of the mSin3A corepressor complex. Mol Cell Biol 23:3456–3467
23. Forler D, Kocher T, Rode M, Gentzel M, Izaurralde E, Wilm M (2003) An efficient protein complex purification method for functional proteomics in higher eukaryotes. Nat Biotechnol 21:89–92

24. Friedel CC, Zimmer R (2009) Identifying the topology of protein complexes from affinity purification assays. Bioinformatics 25:2140–2146
25. Gavin AC, Aloy P, Grandi P, Krause R, Boesche M, Marzioch M, Rau C, Jensen LJ, Bastuck S, Dumpelfeld B et al (2006) Proteome survey reveals modularity of the yeast cell machinery. Nature 440:631–636
26. Greenman C, Stephens P, Smith R, Dalgliesh GL, Hunter C, Bignell G, Davies H, Teague J, Butler A, Stevens C et al (2007) Patterns of somatic mutation in human cancer genomes. Nature 446:153–158
27. Gu W, Malik S, Ito M, Yuan CX, Fondell JD, Zhang X, Martinez E, Qin J, Roeder RG (1999) A novel human SRB/MED-containing cofactor complex, SMCC, involved in transcription regulation. Mol Cell 3:97–108
28. Guenther MG, Lane WS, Fischle W, Verdin E, Lazar MA, Shiekhattar R (2000) A core SMRT corepressor complex containing HDAC3 and TBL1, a WD40-repeat protein linked to deafness. Genes Dev 14:1048–1057
29. Guglielmi B, van Berkum NL, Klapholz B, Bijma T, Boube M, Boschiero C, Bourbon HM, Holstege FC, Werner M (2004) A high resolution protein interaction map of the yeast Mediator complex. Nucleic Acids Res 32:5379–5391
30. Haber DA, Settleman J (2007) Cancer: drivers and passengers. Nature 446:145–146
31. Han JD, Bertin N, Hao T, Goldberg DS, Berriz GF, Zhang LV, Dupuy D, Walhout AJ, Cusick ME, Roth FP et al (2004) Evidence for dynamically organized modularity in the yeast protein-protein interaction network. Nature 430:88–93
32. Hart GT, Lee I, Marcotte ER (2007) A high-accuracy consensus map of yeast protein complexes reveals modular nature of gene essentiality. BMC Bioinformatics 8:236
33. Isserlin R, El-Badrawi RA, Bader GD (2011) The biomolecular interaction network database in PSI-MI 2.5. Database (Oxford) 2011:baq037
34. Junttila MR, Saarinen S, Schmidt T, Kast J, Westermarck J (2005) Single-step Strep-tag purification for the isolation and identification of protein complexes from mammalian cells. Proteomics 5:1199–1203
35. Keshava Prasad TS, Goel R, Kandasamy K, Keerthikumar S, Kumar S, Mathivanan S, Telikicherla D, Raju R, Shafreen B, Venugopal A et al (2009) Human protein reference database–2009 update. Nucleic Acids Res 37:D767–D772
36. Kocher T, Superti-Furga G (2007) Mass spectrometry-based functional proteomics: from molecular machines to protein networks. Nat Methods 4:807–815
37. Krogan NJ, Cagney G, Yu H, Zhong G, Guo X, Ignatchenko A, Li J, Pu S, Datta N, Tikuisis AP et al (2006) Global landscape of protein complexes in the yeast Saccharomyces cerevisiae. Nature 440:637–643
38. Lanz RB, Bulynko Y, Malovannaya A, Labhart P, Wang L, Li W, Qin J, Harper M, O'Malley BW (2010) Global characterization of transcriptional impact of the SRC-3 coregulator. Mol Endocrinol 24:859–872
39. Le Guezennec X, Vermeulen M, Brinkman AB, Hoeijmakers WA, Cohen A, Lasonder E, Stunnenberg HG (2006) MBD2/NuRD and MBD3/NuRD, two distinct complexes with different biochemical and functional properties. Mol Cell Biol 26:843–851
40. Li Y, Bjorklund S, Jiang YW, Kim YJ, Lane WS, Stillman DJ, Kornberg RD (1995) Yeast global transcriptional regulators Sin4 and Rgr1 are components of mediator complex/RNA polymerase II holoenzyme. Proc Natl Acad Sci USA 92:10864–10868
41. Lupien M, Eeckhoute J, Meyer CA, Wang Q, Zhang Y, Li W, Carroll JS, Liu XS, Brown M (2008) FoxA1 translates epigenetic signatures into enhancer-driven lineage-specific transcription. Cell 132:958–970
42. Majka J, Burgers PM (2004) The PCNA-RFC families of DNA clamps and clamp loaders. Prog Nucleic Acid Res Mol Biol 78:227–260
43. Malovannaya A, Lanz RB, Jung SY, Bulynko Y, Le NT, Chan DW, Ding C, Shi Y, Yucer N, Krenciute G, Kim BJ, Li C, Chen R, Li W, Wang Y, O'Malley BW, Qin J (2011) Analysis of the human endogenous coregulator complexome. Cell 145(5):787–799

44. Malovannaya A, Li Y, Bulynko Y, Jung SY, Wang Y, Lanz RB, O'Malley BW, Qin J (2010) Streamlined analysis schema for high-throughput identification of endogenous protein complexes. Proc Natl Acad Sci USA 107:2431–2436
45. Matsuoka S, Ballif BA, Smogorzewska A, McDonald ER 3rd, Hurov KE, Luo J, Bakalarski CE, Zhao Z, Solimini N, Lerenthal Y et al (2007) ATM and ATR substrate analysis reveals extensive protein networks responsive to DNA damage. Science 316:1160–1166
46. Morell M, Espargaro A, Aviles FX, Ventura S (2008) Study and selection of in vivo protein interactions by coupling bimolecular fluorescence complementation and flow cytometry. Nat Protoc 3:22–33
47. Mu JJ, Wang Y, Luo H, Leng M, Zhang J, Yang T, Besusso D, Jung SY, Qin J (2007) A proteomic analysis of ataxia telangiectasia-mutated (ATM)/ATM-Rad3-related (ATR) substrates identifies the ubiquitin-proteasome system as a regulator for DNA damage checkpoints. J Biol Chem 282:17330–17334
48. Nesvizhskii AI (2010) A survey of computational methods and error rate estimation procedures for peptide and protein identification in shotgun proteomics. J Proteomics 73:2092–2123
49. Old WM, Meyer-Arendt K, Aveline-Wolf L, Pierce KG, Mendoza A, Sevinsky JR, Resing KA, Ahn NG (2005) Comparison of label-free methods for quantifying human proteins by shotgun proteomics. Mol Cell Proteomics 4:1487–1502
50. Pagel P, Kovac S, Oesterheld M, Brauner B, Dunger-Kaltenbach I, Frishman G, Montrone C, Mark P, Stumpflen V, Mewes HW et al (2005) The MIPS mammalian protein-protein interaction database. Bioinformatics 21:832–834
51. Paoletti AC, Parmely TJ, Tomomori-Sato C, Sato S, Zhu D, Conaway RC, Conaway JW, Florens L, Washburn MP (2006) Quantitative proteomic analysis of distinct mammalian Mediator complexes using normalized spectral abundance factors. Proc Natl Acad Sci USA 103:18928–18933
52. Park PJ (2009) ChIP-seq: advantages and challenges of a maturing technology. Nat Rev Genet 10:669–680
53. Poser I, Sarov M, Hutchins JR, Heriche JK, Toyoda Y, Pozniakovsky A, Weigl D, Nitzsche A, Hegemann B, Bird AW et al (2008) BAC TransgeneOmics: a high-throughput method for exploration of protein function in mammals. Nat Methods 5:409–415
54. Ranish JA, Yi EC, Leslie DM, Purvine SO, Goodlett DR, Eng J, Aebersold R (2003) The study of macromolecular complexes by quantitative proteomics. Nat Genet 33:349–355
55. Ravasi T, Suzuki H, Cannistraci CV, Katayama S, Bajic VB, Tan K, Akalin A, Schmeier S, Kanamori-Katayama M, Bertin N et al (2010) An atlas of combinatorial transcriptional regulation in mouse and man. Cell 140:744–752
56. Rinner O, Mueller LN, Hubalek M, Muller M, Gstaiger M, Aebersold R (2007) An integrated mass spectrometric and computational framework for the analysis of protein interaction networks. Nat Biotechnol 25:345–352
57. Ruepp A, Waegele B, Lechner M, Brauner B, Dunger-Kaltenbach I, Fobo G, Frishman G, Montrone C, Mewes HW (2010) CORUM: the comprehensive resource of mammalian protein complexes–2009. Nucleic Acids Res 38:D497–D501
58. Ruffner H, Bauer A, Bouwmeester T (2007) Human protein-protein interaction networks and the value for drug discovery. Drug Discov Today 12:709–716
59. Salwinski L, Miller CS, Smith AJ, Pettit FK, Bowie JU, Eisenberg D (2004) The database of interacting proteins: 2004 update. Nucleic Acids Res 32:D449–D451
60. Santarius T, Shipley J, Brewer D, Stratton MR, Cooper CS (2010) A census of amplified and overexpressed human cancer genes. Nat Rev Cancer 10:59–64
61. Sardiu ME, Cai Y, Jin J, Swanson SK, Conaway RC, Conaway JW, Florens L, Washburn MP (2008) Probabilistic assembly of human protein interaction networks from label-free quantitative proteomics. Proc Natl Acad Sci USA 105:1454–1459
62. Sardiu ME, Florens L, Washburn MP (2009) Evaluation of clustering algorithms for protein complex and protein interaction network assembly. J Proteome Res 8:2944–2952

63. Scholtens D, Vidal M, Gentleman R (2005) Local modeling of global interactome networks. Bioinformatics 21:3548–3557
64. Sjoblom T, Jones S, Wood LD, Parsons DW, Lin J, Barber TD, Mandelker D, Leary RJ, Ptak J, Silliman N et al (2006) The consensus coding sequences of human breast and colorectal cancers. Science 314:268–274
65. Sowa ME, Bennett EJ, Gygi SP, Harper JW (2009) Defining the human deubiquitinating enzyme interaction landscape. Cell 138:389–403
66. Steen H, Mann M (2004) The ABC's (and XYZ's) of peptide sequencing. Nat Rev Mol Cell Biol 5:699–711
67. Taatjes DJ, Naar AM, Andel F 3rd, Nogales E, Tjian R (2002) Structure, function, and activator-induced conformations of the CRSP coactivator. Science 295:1058–1062
68. Tackett AJ, DeGrasse JA, Sekedat MD, Oeffinger M, Rout MP, Chait BT (2005) I-DIRT, a general method for distinguishing between specific and nonspecific protein interactions. J Proteome Res 4:1752–1756
69. Varghese JS, Easton DF (2010) Genome-wide association studies in common cancers–what have we learnt? Curr Opin Genet Dev 20:201–209
70. Venkatesan K, Rual JF, Vazquez A, Stelzl U, Lemmens I, Hirozane-Kishikawa T, Hao T, Zenkner M, Xin X, Goh KI et al (2009) An empirical framework for binary interactome mapping. Nat Methods 6:83–90
71. Wang W, Cote J, Xue Y, Zhou S, Khavari PA, Biggar SR, Muchardt C, Kalpana GV, Goff SP, Yaniv M et al (1996) Purification and biochemical heterogeneity of the mammalian SWI-SNF complex. EMBO J 15:5370–5382
72. Wang Y, Cortez D, Yazdi P, Neff N, Elledge SJ, Qin J (2000) BASC, a super complex of BRCA1-associated proteins involved in the recognition and repair of aberrant DNA structures. Genes Dev 14:927–939
73. Wang Y, Zhang H, Chen Y, Sun Y, Yang F, Yu W, Liang J, Sun L, Yang X, Shi L et al (2009) LSD1 is a subunit of the NuRD complex and targets the metastasis programs in breast cancer. Cell 138:660–672
74. Weake VM, Workman JL (2010) Inducible gene expression: diverse regulatory mechanisms. Nat Rev Genet 11:426–437
75. Weston AD, Hood L (2004) Systems biology, proteomics, and the future of health care: toward predictive, preventative, and personalized medicine. J Proteome Res 3:179–196
76. Xue Y, Wong J, Moreno GT, Young MK, Cote J, Wang W (1998) NURD, a novel complex with both ATP-dependent chromatin-remodeling and histone deacetylase activities. Mol Cell 2:851–861

Chapter 9
Network Pharmacology to Aid the Drug Discovery Process

Ruth Dannenfelser, Huilei Xu, Catherine Raimond, and Avi Ma'ayan

Abstract The drug discovery process is expensive and time consuming. Hundreds of drugs enter the research and development pipeline only to be dismissed in late phase due to some risky side effect or lack of efficacy in human subjects. Using existing knowledge of approved drugs we can compare experimental drugs' expression profiles and chemical structures to predict their mechanism of action, filtering out the drugs that will not survive the development process saving time and money. We can also use these approaches in combination with clinical data to repurpose and combine existing drugs for improved therapeutic use. Here we discuss many of the current approaches in network pharmacology which can aid in the drug discovery process. First, we describe the fundamental data that forms the basis of these approaches and investigate where we can find this data. Next, we present how to use different data types incorporating network approaches to model drug effects, including various tools and algorithms developed for this purpose. Finally, we present a global overview of how to apply all of these techniques for accurate side effect and new indication predictions.

Keywords Systems pharmacology • Drug networks • Drug-drug interactions • Adverse events prediction • Gene-phenotype interactions

9.1 Introduction

For the past few decades drug discovery methodologies have focused on the one drug one target paradigm. Derived from the "magic bullet" philosophy, the idea that a drug can be safer and more effective if it is highly selective has been the driving

R. Dannenfelser • H. Xu • C. Raimond • A. Ma'ayan (✉)
Department of Pharmacology and Systems Therapeutics, Systems Biology Center New York (SBCNY), Mount Sinai School of Medicine, New York, NY 10029, USA
e-mail: avi.maayan@mssm.edu

A. Ma'ayan and Ben D. MacArthur (eds.), *New Frontiers of Network Analysis in Systems Biology*, DOI 10.1007/978-94-007-4330-4_9,

force behind the drug development process. Recent studies have invalidated this approach. Single gene knockout studies in mice have shown little effect on phenotype while other studies have shown that effective drugs once thought to be selective, act on multiple targets [1]. These insights are gradually permeating the pharmaceutical industry which has been frustrated due to many recent failures. In the past 6 years the identification of new molecular entities (NMEs) have dropped by 50% [2]. Even worse, in the period from 1999 to 2008 only 17 novel drugs were approved from target-based approaches [3]. Additionally, a high percentage of failures occur in phase III of the development process due to lack of efficacy or detrimental side effects discovered in clinical trials. Network pharmacology offers a new alternative as drug discovery moves forward from the outdated one drug one target methodology. By constructing network models and/or classifiers, network pharmacology can aid in the discovery of new drugs, suggest new indications for approved drugs, and predict potential adverse reactions before they manifest, saving time, money and potentially saving lives.

9.2 Organizing and Integrating Datasets to Make Predictions

The basis of the data surrounding drug-effect predictions is the existing knowledge of currently marketed compounds and their effects on mammalian cells, organs, tissues and the entire patient or model organism. New drugs can be compared with existing drugs to predict their potential benefits and drawbacks. The data to achieve this goal are scattered amongst a plethora of databases as drug-drug, drug-target [4], protein-protein, protein-metabolite, gene regulation, cell signaling [5], drug-phenotype, and gene-phenotype interactions, spanning chemical, pharmacological, biochemical, clinical and genomic domains (Fig. 9.1). We begin by surveying drug-target interaction databases.

9.3 Connecting Drugs to Their Targets

DrugBank [6], the Therapeutic Targets Database (TTD) [7], the Potential Drug Target Database [8], Drug Combination database (DCDB) [9], SuperTarget as well as Matador [10] and the tropical disease research (TDR) Targets [11] are some of the most popular drug-target databases. DrugBank is arguably the most comprehensive resource available to network pharmacology researchers. DrugBank contains extensive data on the nomenclature, ontology, chemistry, structure, function, action, pharmacology, pharmacokinetics, metabolism and pharmaceutical properties of drugs in addition to their primary targets. TTD provides information about the primary targets of approved drugs, drugs in clinical trials, and investigational drugs. TTD is useful when searching for the similarity between drugs and targets, and also for understanding the mechanisms of multi-target drugs. The Potential Drug Target Database provides possible drug targets using the 3D

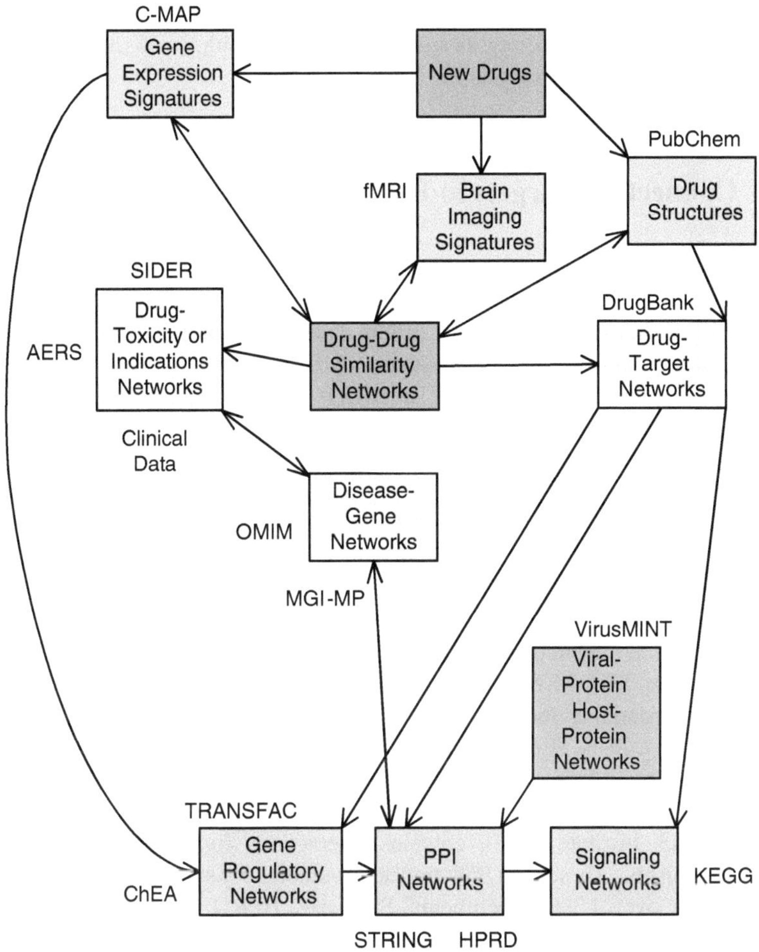

Fig. 9.1 Pipeline for predicting side-effects and indications for new drugs. Networks and data types are listed in *colored rectangles*. Databases and resources that contain these data types are listed near *rectangles*. *Arrows* indicate information flow

structures of targets from the Protein Data Bank, annotated by therapeutic area and associated diseases. DCDB collects drug combination information, including combined activity and indications from more than 6,000 references. Furthermore, DCDB provides known molecular targets, target sequence and structure, and target participated pathways for each individual drug. Therefore, DCDB can facilitate our understanding of the mechanistic interactions between drugs based on their shared targets and pathways. SuperTarget contains data for over 700 drug-target pairs derived from literature mining with querying capabilities for drugs, targets, pathways, ontologies or cytochrome P450 enzymes; whereas Matador is a special subset of the drug-target pairs from SuperTarget that are well annotated, useful as a training set for various classifiers. Another interesting more focused drug-target database is the tropical disease research (TDR) Targets. TDR Targets contains data on lesser studied tropical diseases and provides users with a web interface to query ranked lists of druggable targets for a given disease.

9.4 Connecting Targets to Pathways and Networks

Drug-target networks can be connected to the growing resources of mammalian protein-protein interaction networks, connecting groups of targets to functional modules of the human interactome. Leading databases of protein-protein interactions include Molecular INTeraction database (MINT) [12], the Biological General Repository for Interaction Datasets (BioGRID) [13], the Human Protein Reference Database (HPRD) [14], the protein interaction database (IntACT) [15], InnateDB [16] and Search Tool for the Retrieval of Interacting Genes (STRING) [17]. The Molecular INTeraction database, MINT contains approximately 200,000 experimentally validated protein-protein interactions from five different species derived from literature. MINT's web interface features the ability to visualize an interaction network for a query set of proteins. The MINT database also contains specialized host-virus protein-protein interactions in VirusMINT, collecting over 5,000 unique interactions between many viruses and human proteins [18]. Since some drugs interact with the life cycle of many viruses, VirusMINT, although currently underutilized, is potentially invaluable for understanding the interplay between drug-induced resurgence of certain viruses as side effects. BioGRID contains over 300,000 unique genetic and protein interactions derived from over 25,000 literature sources for human with additional unique interactions for other model organisms. IntAct, created by the European Bioinformatics Institute (EBI), contains interactions for human and four other model organisms. These 154,958 molecular interactions in IntAct are manually curated from over 4,300 experiments. Similarly, HPRD is a high quality source of literature-based protein-protein interaction data from low-throughput studies. It also contains a variety of protein specific information, such as post translational modifications, enzyme-substrate relationships and disease associations. HPRD also provides a web interface with tools such as PhosophoMotif Finder which searches a protein for potential phosphorylation motif sites described in literature, a signaling pathway resource called NetPath, and the ability to perform BLAST searches on protein or nucleotide sequences. These protein-protein interaction databases are few of the major resources for this type of data. Some additional resources such as InnateDB [16] consolidate interactions from many of the available databases as well as develop their own focused interactomes. Finally, the STRING database, probably the most comprehensive database, consolidates many different types of interactions, including protein-protein interactions. The database contains confidence scores for interactions based on aggregated evidence and is being widely used by the systems biology community. Regardless, many more academic and commercial mammalian protein-protein interaction databases exist and continually appear. For a comprehensive list of them readers may refer to PathGuide [19] an effort for collecting and indexing such datasets.

Additional information from pathway databases and/or transcriptional regulation networks can be easily integrated with protein-protein interaction networks adding another layer toward understanding the global effects of drugs on human cells. One of the major resources for systems level regulatory information is the Kyoto Encyclopedia of Genes and Genomes (KEGG) [20]. KEGG contains 16 major databases divided into 3 categories: systems, genomic, and chemical information. The KEGG pathway database is especially useful for its rich information on metabolic and cellular processes. Such datasets from KEGG can be easily integrated into interaction networks from other sources. Databases such as UniPROBE [21], JASPAR [22], TRANSFAC [23], and ChEA [24] provide regulatory relationships between transcription factors and their putative target genes. UniPROBE, JASPAR and TRANSFAC are similar databases that primarily store transcription factor binding sites and allow for the utilization of sequence logo motifs, also called position weight matrices. ChIP Enrichment Analysis (ChEA), is a database with a web based user interface storing transcription factor-target gene pairs derived from ChIP-chip and ChIP-seq experiments. Users of ChEA can input a list of genes and view a ranked list of their predicted transcription factor regulators. Similarly to ChEA, kinase enrichment analysis (KEA) [25] is a database with a web interface storing kinase-substrate interactions from literature. This tool allows users to connect lists of proteins to the protein kinases that are known to phosphorylate members of the input list based on prior literature reports. There are many other types of regulatory networks of mammalian cells that can be integrated with protein-protein interactions networks and drug-target networks. Here we just scratch the surface, listing a short subset of these resources with the aim of later explaining how these specific resources can be used to build network models and classifiers to predict drug mechanisms of action, side effects and new indications.

9.5 Making Predictions Without Considering the Human Interactome: Drug-Drug Similarity Networks

While drug-target interactions can be connected to protein-protein networks, transcription-factor target gene regulatory networks, metabolic or cell signaling networks, these networks are still incomplete. Moreover, these networks contain research focus biases, as well as many false positive and false negatives. The question that we can ask then is: "can we predict side effects and new indications for drugs without considering their targeted networks?" The short answer is yes. One way to do this is to construct drug-drug similarity networks. By constructing drug-drug similarity networks Gottlieb et al. demonstrated how these could be used to accurately predict indications [26]. Networks of drug-drug similarity can be generated from different sources where the edges connecting pairs of drugs represent different aspects of drug-drug similarity. The structure of chemical compounds is often used in the early phase of the drug development cycle to identify candidate compounds, making it a good first choice to consider. If two drugs share similarities within their structure, it is likely they share side effects and indications.

Keiser et al. [27] showed that examining the similarity of the structure of compounds can aid in the characterization of off-target effects. PubChem [28], a database developed by NCBI, stores entries of chemical molecules and their activities against various biological assays. Users of PubChem can search for various compound structures using a name, SMILES string, CID, or InChI identifiers and find similar compounds from one of three databases containing compounds, substances and bioassay information. The commercial correlate to PubChem is the American Chemical Society's Chemical Abstract Service (CAS). As the leading source of chemical information, CAS contains data on over 50 million substances and 60 million protein and DNA sequences. DrugBank can also be a useful source for building drug similarity networks, for example, by using the SMILES strings for all FDA approved drugs. Other commercial databases such as the MDL Drug Data Report (MDDR) [29] and the World of Molecular Bioactivity (WOMBAT) [30] can be used to build drug similarity networks. Once we construct drug-drug networks based on structure, network diffusion kernel classifiers can be used for predicting side effects and indications for new drugs.

In addition to using the chemical structure of a compound to compare drug pairs and construct drug-drug networks, we can also use gene expression signatures, generated by applying different drugs to the same cell lines and measuring changes in genome-wide expression. Hu and Agarwal [31] did exactly that. They created disease-disease, disease-drug, and drug-drug networks from expression signatures. By comparing drug induced expression profiles, we may be able to predict potential toxicity and efficacy. The main resource for gene expression profiles is the Gene Expression Omnibus (GEO) a database curated by NCBI [32]. GEO is a public repository of functional genomics data from microarrays, next-generation sequencing, and other high-throughput experiments. Users can search and browse GEO's web interface and download profiles of interest. The Connectivity Map (CMAP) database created by researchers at the Broad Institute is a central resource for drug induced expression profiles [33]. CMAP contains drug signatures for almost all FDA approved small molecule compounds in well established cell lines, usually under a single exposure, and in most cases covering one dose. The CMAP approach provides a new avenue, a "signature-based" approach, for discovering new drugs, predicting side effects and discovering new indications. Such an approach is not target centered but molecular signature centered. The goal is to find novel small molecules that induce a gene expression signature that can correlate to a desired phenotype, i.e., a clinical outcome. Known drugs can be compared based on their gene expression signatures to predict new expected indications as well as forecast side effects based on signature similarity. The signature itself should not just be used as an indicator vector that contains potential information about drug similarity. It has information that can be used to better understand which pathways are perturbed by the drugs, linking changes in expression back to the targets. Recent analysis of CMAP showed that human cells tend to either increase or decrease the mRNA expression of their known drug targets. This insight suggests that drug induced gene expression signatures such as the data in CMAP can be used to recover drug targets and drug targeted pathways for drugs where this information is not known [34]. The study also points out that the data in CMAP suffers from the "batch effect", showing similarity between pairs of drugs not due to the drug effect but due to biases in the experimental setup. Another general limitation of the CMAP approach is that different cell types may respond differently to different drugs. In addition, changes in expression at one time point may mask similarity between drugs that show similar effects at different time points. Such challenges are currently being addressed by developing technologies that can obtain signatures at lower cost with higher throughput [35]. Regardless of these few limitations, the CMAP approach is a breath of fresh air for the drug discovery process, expected to revolutionize the field. Once more widely accepted, target-based drug discovery will take a lesser role to the emerging signature-based approach. Besides building drug-drug similarity networks based on gene expression signatures, the signature-based approach can be used for personalized medicine. Tissues from patients can be profiled with expression arrays to stratify patients, and drug treatments can be tailored to individuals based on how cells with similar expression profile respond to drugs in a dish. One of the limitations of the drug signature-based approach applied to profile tissue from patients is that it would be difficult to assess the effects of drugs on different additional tissues including the brain. Similarly to the drug induced gene expression molecular signatures used to build drug-drug similarity networks, we can envision that in the future technologies such as functional Magnetic Resonance Imaging (fMRI) may be useful for assessing toxicity, efficacy, and determining dosage for drugs affecting the human brain by building drug induced profiles of brain activity [36]. fMRI profiling could be used in conjunction with drugs to obtain drug signatures for different patients based on drug induced changes in brain activity.

9.6 Linking Genes to Phenotype

On the other "side" of the inputs, the drugs, and their direct molecular effects on cellular networks, are the outputs: the phenotypes, the clinical outcomes, the diseases, and the side effects or adverse events caused by the drugs. These phenotypic outcomes can be linked to each other based on different measures of similarity, or linked to individual genes through genetic studies. Let's begin with describing some sources for obtaining information about side effects: The authoritative resource for drug side effects is SIDER, the computer readable side effect resource. SIDER contains 888 FDA approved drugs connected to 1,450 side effects extracted from literature and package inserts. In addition to drug-side effect pairs, SIDER contains the frequency of which these adverse events occur [37]. The adverse events reporting system (AERS), although a less definitive resource can be used to infer additional drug-side effect relations. Developed by the FDA, AERS is a database derived from spontaneous reports written by physicians listing side effects and all of the drugs taken by the patient when the side effect occurred.

A drug/adverse-event pair in AERS does not guarantee a causal relationship between the drug and the side effect, but may suggest one. The data in AERS also lack the proper controls, for example, knowing how many patients took the drug but did not display the side effect. This is also known as the "denominator problem".

A useful resource for phenotypic information is the Mouse Genome Informatics Mouse Phenotype MGI-MP ontology [38]. Developed by The Jackson Laboratory, the MGI-MP browser links phenotypes for knockout mice to single gene knockouts. Such phenotypes can be mapped to side effects and diseases observed in humans. The MGI-MP database links genes to phenotypes through genetic animal studies. Similarly, associated genes directly linked to human disease, but with fewer genes, can be found at the Online Mendelian Inheritance in Man (OMIM) morbid map. OMIM is the authoritative resource for listing diseases that have known disease genes where genetic studies showed that alteration to the normal genomic sequence for the gene resulted in the disease phenotype [38]. The database primarily covers genetic disorders but there are other diseases that can be considered. In addition, recent advances such as those from genome-wide association studies (GWAS), whole exome sequencing, and whole genome sequencing [39] are expected to expand our knowledge of disease genes rapidly.

Organizing diseases based on their similarities is another step toward making the network pharmacology revolution a reality. Diseases and phenotype ontologies are rapidly developed for addressing this aspect of the puzzle [40–42]. Another way to go about organizing diseases, side effects and phenotypes is building networks of disease-disease similarity. Rzhetsky et al. [43] and Barabási and colleagues [44] were among the first to connect diseases based on shared disease genes. Many other similar studies followed.

Traditionally, GWAS and linkage mapping are the key techniques for producing candidate disease genes. However, linking these candidate genes with their causal mutations remains a challenge [45]. One of the simplest techniques is to examine the network neighborhoods around the genes known to be associated with a particular disease phenotype. This strategy is derived from the observation that mutations near neighboring disease genes were found to cause similar phenotypes [46]. Many studies capitalized on this observation, predicting novel disease genes given few known disease genes [47–53]. So far we have only mentioned drug target and disease gene prediction using the human protein-protein interaction network but we can also incorporate expression data or other types of datasets to identify functional relationships between genes to build gene-gene or protein-protein networks. For example, by examining the co-expression of genes across cell types and tissues, we could potentially identify novel disease genes given known seed disease genes based on co-expression [54]. Hence, drug-drug, drug-target, protein-protein interactions, metabolic and gene regulation networks can be linked to phenotypes, diseases and side effects through the genes they share. We can start from drugs, the inputs, to diseases and side effects, the outputs, through genes and their surrounding networks identifying patterns and modules that can be useful for making predictions for side effects and indications as well as unravel novel molecular mechanisms of diseases and drugs.

9.7 Summary

In conclusion, we can develop a pipeline (Fig. 9.1) that starts with new candidate compounds. Knowing the drug structure we can compare the new drug to already approved drugs to deduce side effects and indications. We can also compare the drug induced gene expression profiles, or the drug effects on the brain, using for example fMRI imaging applied in a standard protocol. If we know the targets for the drug, we can build a network neighborhood around the targets using protein-protein interactions or other types of gene-gene relationship networks. If the targets fall around a disease module this might lead to a desired indication or an undesired side effect. If we do not know the targets or wish to identify the pathways affected by the drug regardless of whether the intended targets are known, we could infer the pathways affected by the drug using gene expression data and tracing back the pathways by first identifying the upstream transcription factors responsible for the changes in expression induced by the drug. We can then connect these transcription factors using protein-protein interactions to build transcriptional complexes centered on the transcription factors that likely regulate the genes that changed in expression due to the drug effect. These transcriptional centered complexes may correctly hint at the pathways perturbed by the drug and may contain known disease genes in which mutations in those genes, or knockout mouse models targeting those genes, are known to lead to undesired phenotypes. Such network-based approaches might help us to determine side effects before they occur, predict new indications for already approved drugs, and attempt to understand how combinations of drugs might work. Hence drug-drug, drug-target, protein-protein, protein-ligand, metabolic, gene regulatory, drug-phenotype, and gene-phenotype interactions can all fit into a global framework that can be used to predict side effects and new indications for new drugs and currently marketed drugs.

Acknowledgements This work was supported by NIH grants P50GM071558-03, R01DK088541-01A1, RC2LM010994-01, P01DK056492-10, RC4DK090860-01, and KL2RR029885-0109 to AM.

References

1. Hopkins AL (2008) Network pharmacology: the next paradigm in drug discovery. Nat Chem Biol 4(11):682–690
2. Paul SM, Mytelka DS, Dunwiddie CT, Persinger CC, Munos BH, Lindborg SR, Schacht AL (2010) How to improve R&D productivity: the pharmaceutical industry's grand challenge. Nat Rev Drug Discov 9(3):203–214
3. Swinney DC, Anthony J (2011) How were new medicines discovered? Nat Rev Drug Discov 10(7):507–519
4. Ma'ayan A, Jenkins SL, Goldfarb J, Iyengar R (2007) Network analysis of FDA approved drugs and their targets. Mt Sinai J Med J Transl Personal Med 74(1):27–32

5. Ma'ayan A, Jenkins SL, Neves S, Hasseldine A, Grace E, Dubin-Thaler B, Eungdamrong NJ, Weng G, Ram PT, Rice JJ et al (2005) Formation of regulatory patterns during signal propagation in a mammalian cellular network. Science 309(5737):1078–1083
6. Wishart DS, Knox C, Guo AC, Cheng D, Shrivastava S, Tzur D, Gautam B, Hassanali M (2008) DrugBank: a knowledgebase for drugs, drug actions and drug targets. Nucleic Acids Res 36 (Suppl 1):D901–D906
7. Chen X, Ji ZL, Chen YZ (2002) TTD: Therapeutic Target Database. Nucleic Acids Res 30 (1):412–415
8. Klipp E, Wade RC, Kummer U (2010) Biochemical network-based drug-target prediction. Curr Opin Biotechnol 21(4):511–516
9. Liu Y, Hu B, Fu C, Chen X (2010) DCDB: drug combination database. Bioinformatics 26 (4):587–588
10. Ganther S, Kuhn M, Dunkel M, Campillos M, Senger C, Petsalaki E, Ahmed J, Urdiales EG, Gewiess A, Jensen LJ et al (2008) SuperTarget and Matador: resources for exploring drug-target relationships. Nucleic Acids Res 36(Suppl 1):D919–D922
11. Aguero F, Al-Lazikani B, Aslett M, Berriman M, Buckner FS, Campbell RK, Carmona S, Carruthers IM, Chan AWE, Chen F et al (2008) Genomic-scale prioritization of drug targets: the TDR Targets database. Nat Rev Drug Discov 7(11):900–907
12. Chatr-aryamontri A, Ceol A, Palazzi LM, Nardelli G, Schneider MV, Castagnoli L, Cesareni G (2007) MINT: the Molecular INTeraction database. Nucleic Acids Res 35 (Suppl 1):D572–D574
13. Stark C, Breitkreutz B-J, Reguly T, Boucher L, Breitkreutz A, Tyers M (2006) BioGRID: a general repository for interaction datasets. Nucleic Acids Res 34(Suppl 1):D535–D539
14. Prasad TSK, Goel R, Kandasamy K, Keerthikumar S, Kumar S, Mathivanan S, Telikicherla D, Raju R, Shafreen B, Venugopal A et al (2009) Human protein reference database-2009 update. Nucleic Acids Res 37(Suppl 1):D767–D772
15. Aranda B, Achuthan P, Alam-Faruque Y, Armean I, Bridge A, Derow C, Feuermann M, Ghanbarian AT, Kerrien S, Khadake J et al (2010) The IntAct molecular interaction database in 2010. Nucleic Acids Res 38(Suppl 1):D525–D531
16. Lynn DJ, Winsor GL, Chan C, Richard N, Laird MR, Barsky A, Gardy JL, Roche FM, Chan THW, Shah N et al (2008) InnateDB: facilitating systems-level analyses of the mammalian innate immune response. Mol Syst Biol 4:218
17. Szklarczyk D, Franceschini A, Kuhn M, Simonovic M, Roth A, Minguez P, Doerks T, Stark M, Muller J, Bork P et al (2011) The STRING database in 2011: functional interaction networks of proteins, globally integrated and scored. Nucleic Acids Res 39(Suppl 1): D561–D568
18. Chatr-aryamontri A, Ceol A, Peluso D, Nardozza A, Panni S, Sacco F, Tinti M, Smolyar A, Castagnoli L, Vidal M et al (2009) VirusMINT: a viral protein interaction database. Nucleic Acids Res 37(Suppl 1):D669–D673
19. Bader GD, Cary MP, Sander C (2006) Pathguide: a pathway resource list. Nucleic Acids Res 34(Suppl 1):D504–D506
20. Kanehisa M, Araki M, Goto S, Hattori M, Hirakawa M, Itoh M, Katayama T, Kawashima S, Okuda S, Tokimatsu T et al (2008) KEGG for linking genomes to life and the environment. Nucleic Acids Res 36(Suppl 1):D480–D484
21. Newburger DE, Bulyk ML (2009) UniPROBE: an online database of protein binding microarray data on protein-DNA interactions. Nucleic Acids Res 37(Suppl 1):D77–D82
22. Sandelin A, Alkema W, Engstrom P, Wasserman WW, Lenhard B (2004) JASPAR: an open-access database for eukaryotic transcription factor binding profiles. Nucleic Acids Res 32 (Suppl 1):D91–D94. doi:D91
23. Matys V, Kel-Margoulis OV, Fricke E, Liebich I, Land S, Barre-Dirrie A, Reuter I, Chekmenev D, Krull M, Hornischer K et al (2006) TRANSFAC and its module TRANSCompel: transcriptional gene regulation in eukaryotes. Nucleic Acids Res 34(Suppl 1):D108–D110

24. Lachmann A, Xu H, Krishnan J, Berger SI, Mazloom AR, Ma'ayan A (2010) ChEA: transcription factor regulation inferred from integrating genome-wide ChIP-X experiments. Bioinformatics 26(19):2438–2444
25. Lachmann A, Ma'ayan A (2009) KEA: kinase enrichment analysis. Bioinformatics 25 (5):684–686
26. Gottlieb A, Stein GY, Ruppin E, Sharan R (2011) PREDICT: a method for inferring novel drug indications with application to personalized medicine. Mol Syst Biol 7:496
27. Keiser MJ, Setola V, Irwin JJ, Laggner C, Abbas AI, Hufeisen SJ, Jensen NH, Kuijer MB, Matos RC, Tran TB et al (2009) Predicting new molecular targets for known drugs. Nature 462 (7270):175–181
28. Wang Y, Xiao J, Suzek TO, Zhang J, Wang J, Bryant SH (2009) PubChem: a public information system for analyzing bioactivities of small molecules. Nucleic Acids Res 37 (suppl 2):W623–W633
29. Schuffenhauer A, Floersheim P, Acklin P, Jacoby E (2002) Similarity metrics for ligands reflecting the similarity of the target proteins. J Chem Inf Comput Sci 43(2):391–405
30. Olah M, Mracec M, Ostopovici L, Rad R, Bora A, Hadaruga N, Olah I, Banda M, Simon Z, Mracec M et al (2005) WOMBAT: world of molecular bioactivity. In: Opera TI (ed) Chemoinformatics in drug discovery. Wiley-VCH Verlag GmbH & Co. KGaA, Weinheim, pp 221–239
31. Hu G, Agarwal P (2009) Human disease-drug network based on genomic expression profiles. PLoS One 4(8):e6536
32. Barrett T, Troup DB, Wilhite SE, Ledoux P, Rudnev D, Evangelista C, Kim IF, Soboleva A, Tomashevsky M, Edgar R (2007) NCBI GEO: mining tens of millions of expression profiles-database and tools update. Nucleic Acids Res 35(Suppl 1):D760–D765
33. Lamb J, Crawford ED, Peck D, Modell JW, Blat IC, Wrobel MJ, Lerner J, Brunet J-P, Subramanian A, Ross KN et al (2006) The connectivity map: using gene-expression signatures to connect small molecules, genes, and disease. Science 313(5795):1929–1935
34. Iskar M, Campillos M, Kuhn M, Jensen LJ, van Noort V, Bork P (2010) Drug-induced regulation of target expression. PLoS Comput Biol 6(9):e1000925
35. Peck D, Crawford E, Ross K, Stegmaier K, Golub T, Lamb J (2006) A method for high-throughput gene expression signature analysis. Genome Biol 7(7):R61
36. Wise RG, Tracey I (2006) The role of fMRI in drug discovery. J Magn Reson Imaging 23 (6):862–876
37. Kuhn M, Campillos M, Letunic I, Jensen LJ, Bork P (2010) A side effect resource to capture phenotypic effects of drugs. Mol Syst Biol 6:343
38. Smith C, Goldsmith C-A, Eppig J (2004) The mammalian phenotype ontology as a tool for annotating, analyzing and comparing phenotypic information. Genome Biol 6(1):R7
39. Gravel S, Henn BM, Gutenkunst RN, Indap AR, Marth GT, Clark AG, Yu F, Gibbs RA, Project TG, Bustamante CD (2011) Demographic history and rare allele sharing among human populations. Proc Natl Acad Sci USA 108(29):11983–11988
40. Osborne J, Flatow J, Holko M, Lin S, Kibbe W, Zhu L, Danila M, Feng G, Chisholm R (2009) Annotating the human genome with Disease Ontology. BMC Genomics 10(Suppl 1):S6
41. Smith CL, Eppig JT (2009) The mammalian phenotype ontology: enabling robust annotation and comparative analysis. Wiley Interdiscip Rev Syst Biol Med 1(3):390–399
42. Gkoutos G, Green E, Mallon A-M, Hancock J, Davidson D (2004) Using ontologies to describe mouse phenotypes. Genome Biol 6(1):R8
43. Rzhetsky A, Wajngurt D, Park N, Zheng T (2007) Probing genetic overlap among complex human phenotypes. Proc Natl Acad Sci USA 104(28):11694–11699
44. Goh K-I, Cusick ME, Valle D, Childs B, Vidal M, Barabási A-L (2007) The human disease network. Proc Natl Acad Sci USA 104(21):8685–8690
45. Barabási A-L, Gulbahce N, Loscalzo J (2011) Network medicine: a network-based approach to human disease. Nat Rev Genet 12(1):56–68

46. Gandhi TKB, Zhong J, Mathivanan S, Karthick L, Chandrika KN, Mohan SS, Sharma S, Pinkert S, Nagaraju S, Periaswamy B et al (2006) Analysis of the human protein interactome and comparison with yeast, worm and fly interaction datasets. Nat Genet 38(3):285–293
47. Oti M, Snel B, Huynen MA, Brunner HG (2006) Predicting disease genes using protein-protein interactions. J Med Genet 43(8):691–698
48. Cordeddu V, Di Schiavi E, Pennacchio LA, Ma'ayan A, Sarkozy A, Fodale V, Cecchetti S, Cardinale A, Martin J, Schackwitz W et al (2009) Mutation of SHOC2 promotes aberrant protein N-myristoylation and causes Noonan-like syndrome with loose anagen hair. Nat Genet 41(9):1022–1026
49. Berger SI, Ma'ayan A, Iyengar R (2010) Systems pharmacology of arrhythmias. Sci Signal 3(118):ra30
50. Krauthammer M, Kaufmann CA, Gilliam TC, Rzhetsky A (2004) Molecular triangulation: bridging linkage and molecular-network information for identifying candidate genes in Alzheimer's disease. Proc Natl Acad Sci USA 101(42):15148–15153
51. Iossifov I, Zheng T, Baron M, Gilliam TC, Rzhetsky A (2008) Genetic-linkage mapping of complex hereditary disorders to a whole-genome molecular-interaction network. Genome Res 18(7):1150–1162
52. Köhler S, Bauer S, Horn D, Robinson PN (2008) Walking the interactome for prioritization of candidate disease genes. Am J Hum Genet 82(4):949–958
53. Vanunu O, Magger O, Ruppin E, Shlomi T, Sharan R (2010) Associating genes and protein complexes with disease via network propagation. PLoS Comput Biol 6(1):e1000641
54. Baggs JE, Hughes ME, Hogenesch JB (2010) The network as the target. Wiley Interdiscip Rev Syst Biol Med 2(2):127–133

Chapter 10
What Can Networks Do for You?

Bernhard M. Schuldt, Franz-Josef Müller, and Andreas A. Schuppert

The first principle [of science] is that you must not fool yourself, and you are the easiest person to fool . . . After you've not fooled yourself, it's easy not to fool other scientists. You just have to be honest in a conventional way after that.

Richard Feynman in his lecture "What is and What Should be the Role of Scientific Culture in Modern Society", given at the Galileo Symposium in Italy (1964).

Abstract This chapter aims at demonstrating the utility of network approaches in classification and outlier detection tasks in the context of stem cell biology and related fields. With modern high-through-put methods it has now become easier and cheaper to accurately measure thousands of features on a genome-wide scale than to define a low number of markers that can be tested, for example with low throughput RT-PCR assays. Typically the number of potential markers exceeds the number of experiments by several orders of magnitude. Therefore the significance – let alone mechanistic involvement – of each possible feature cannot be guaranteed from the data alone. Fortunately, easy-to-use implementations of many powerful network based algorithms have been made freely available so one can readily employ these advanced algorithms on new high-content datasets.

B.M. Schuldt (✉) • A.A. Schuppert
Aachen Institute for Advanced Study in Computational Engineering Science, Rheinisch-Westfälische Technische Hochschule Aachen, Aachen, Germany
e-mail: schuldt@aices.rwth-aachen.de

F.-J. Müller
Department for Psychiatry, Zentrum für Integrative Psychiatrie, Kiel, Germany

A. Ma'ayan and Ben D. MacArthur (eds.), *New Frontiers of Network Analysis in Systems Biology*, DOI 10.1007/978-94-007-4330-4_10,

We will exemplify how network information and structure can be used to improve the prediction of biological phenotypes, and discuss methodological considerations pertinent to enabling reliable and biologically meaningful inferences from *in silico* network studies. We will touch upon difficulties inferring "true" (i.e. mechanistic) networks from biological data and note that, from a practical standpoint, *in silico* networks need not to fully reflect observable biological phenomena for real-world predictability and utility. We have found that a particularly successful strategy is to use statistical learning theory as a stringent framework for comparative evaluation of alternative network methods. This pragmatic and evolutionary approach can be adopted in several biological realms and makes optimal use of todays sophisticated network modeling methodologies. We observe that only such a rigorous workflow can guarantee reproducibility of network-based findings.

Keywords Networks • Pluripotent stem cells • Differentiation • Machine learning • Biomarker • Network modeling

10.1 Current Challenges in Stem Cell Classification

We will discuss three primary current challenges to stem cell classification problems:

1. The insufficiency of classification by "stemness" markers,
2. Low signal-to-noise ratios in heterogeneous cell populations,
3. Identification of hidden or latent variables.

Microarray-based stem cell classification aims at predicting the phenotype of a biological sample based on patterns observed in a training dataset, which was used for "learning" the classifier [1]. Conventional cell biology classification approaches rely on biomarkers – molecules with a known relation to the phenotype, for instance the transcription factor POU5F1/OCT4 or the cell surface glycoprotein TRA1-81 for pluripotent stem cells (PSC). However, in practice this concept turns out to be highly problematic as such stem cell "markers" might not be sufficiently specific and sensitive. Most transcription factors such as OCT4 are expressed in families of cell-types. OCT4 was initially identified as expressed in oocytes and at several stages of pre- and post-implantation embryonic development [2, 3] and OCT4 may be used to induce pluripotency in somatic cells [4, 5]. However, the mere expression of OCT4 on the mRNA or protein level is not indicative of successful reprogramming and other functional markers, such as the activation of the endogenous OCT4 gene, is often used to demonstrate pluripotency-associated epigenetic reprogramming towards a pluripotent state [4]. TRA1-81, on the other hand, was initially identified as marker for teratocarcinoma cells [6] and its expression on the surface of genetically normal pluripotent stem cells reminds us of the close relationship of embryonic stem cells (ESCs) with teratocarcinoma cell lines.

While such challenges might be surmounted by clever experimental designs, measuring stem cell marker signals with satisfactory predictive accuracy can be problematic. For instance, expression levels of genes causally related to a cellular phenotype might be very low, or only detectable in small subpopulations of cells. In this case, gene expression measurements may be not well differentiated from background noise. However, considering the context provided by neighboring genes in an underlying regulatory network allows to "borrow" signal strength and significance. Since measurement noise is generally assumed to be randomly distributed across all measurements, its effect on the average of such connected groups of markers is mitigated (see Fig. 10.1a, b).

Even more problematic are co-varying biological effects caused by unmeasured, undocumented and therefore un-modeled effects. Most prominent are lab and batch effects [7]. Since every researcher and each group employs slightly, but unfortunately consistently, different culture strategies and microarray processing procedures it can be difficult to distinguish genuine phenotypes from irrelevant technical and biological variations. These effects can be biological and/or technical (Fig. 10.1) and can result in signals detectable in sub-networks with biological interpretation. For instance there is an ongoing debate whether the observed differences between induced pluripotent stem cells (iPSCs) and ESCs can be explained by lab-effects or if there is a systematic biological difference that might, or might not, affect the viability of iPSCs in clinical research [8–10]. A more diverse sample collection and a consistent, detailed annotation of experimental procedures and conditions can help to identify co-varying features that have little biological meaning but might profoundly confound high-content data analyses [1, 11].

Finally, it is often the case that the variables that are the real "drivers" of a biological phenomena, may be unavailable to direct measurement, either because they are genuinely hard to measure or because standard methods do not contain functional probes for these factors. Any model will neglect important variables (frequently referred to as "latent variables"). Fortunately, a model that describes the behavior of only a subset of all variables of the cell will be predictive if the measured subset contains full information with respect to the cell state, and the latent variables are linked to the identified cellular phenotype [12]. Even a distinct and unique cellular phenotype might allow for significantly variable biological behavior among representative cell lines. For example, if we are interested in the propensity of a specific pluripotent cell-type to differentiate, for instance along the neural linage, this might be effected by the genetic background of the cell; specific epigenetic marks [11, 13]; the mix of growth factors and differentiation factors in the culture media [14, 15] and the relative influence of pluripotency factors [16]; growth and differentiation-related modules such as microRNAs [17] or mitochondrial gene networks in the cell. Hence, we have to identify and integrate heterogeneous variables representing the effects of all these factors on cell fate into a single model in an appropriate manner, without being ever able to account for all of them. This is a common challenge in mathematical modeling of chemical and industrial processes. The notions of controllability and "observability" of a system without perfect knowledge go back to the work of Kalman

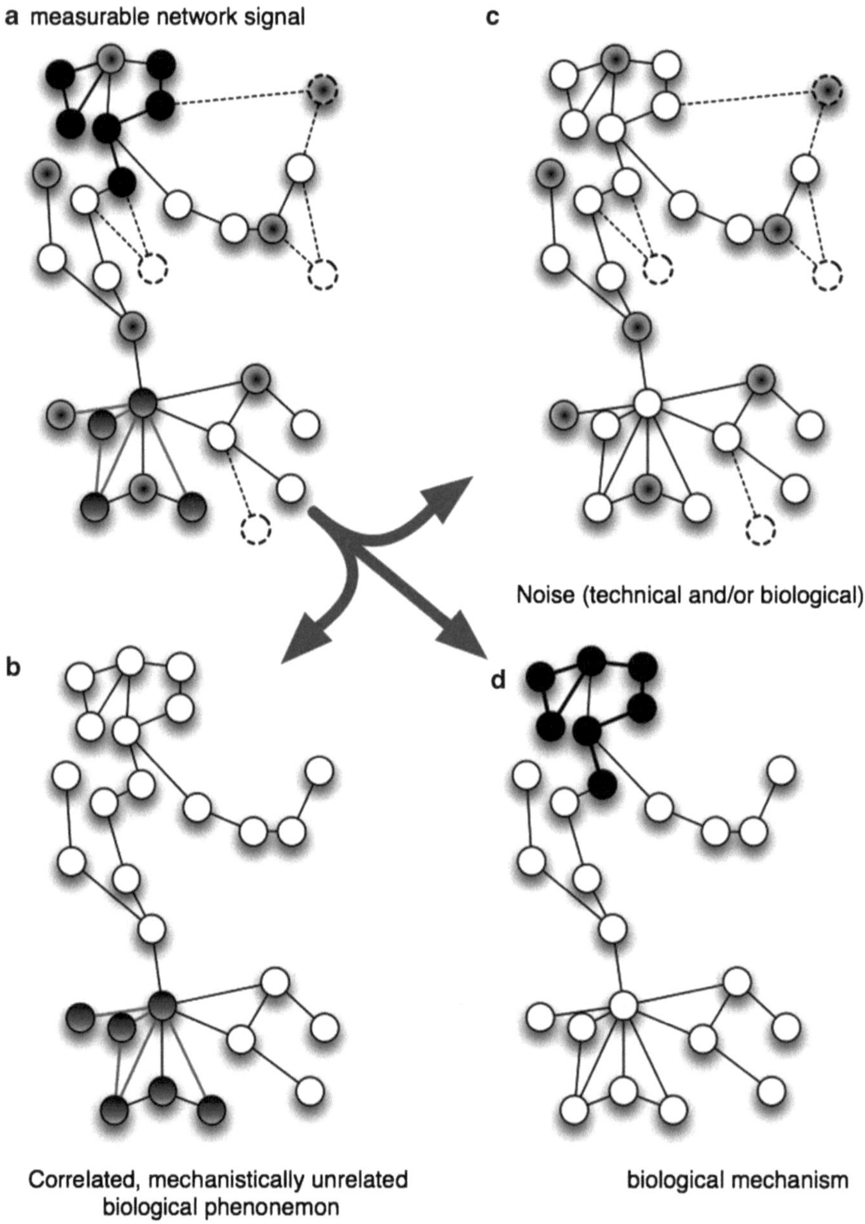

Fig. 10.1 Source separation in network modeling tasks. A measurable network signal that can be gleaned from, for example, transcriptome data and protein-protein interaction or transcription factor-DNA interaction data. Such a measured and/or inferred network (**a**) will hopefully contain the studied biological signal (**d**) but also co-varying biological phenomena (**c**) and technical noise distributed stochastically over the whole network (**b**, *shaded nodes*) as well as false positive links and nodes (**b**, nodes and edges with *dashed lines*). For meaningful biological inferences, the different signal sources will need to be separated and the biological mechanism isolated

and mark the beginning of modern control theory [18]. These mathematical tools are increasingly used in the theoretical and applied study of complex networks [19].

It may come to a surprise to many scientists trained in wet lab biology that these inferred variables need not to play a significant causal role in the mechanism to be modeled: it is sufficient that they allow the (indirect) observation of cell fate. To illustrate this concept consider that the number of ashtrays is closely associated with the number of deaths from lung cancer, thus might be a suitable parameter for modeling lung cancer incidence in diverse populations. Yet, ashtrays are probably not necessarily causing lung cancer. This "observer concept" is well established and has been mathematically analyzed in process engineering.

Networks can build a bridge between observable variables without direct mechanistic relevance and "hidden" variables, which could be driving a specific biological phenomenon. For example networks can be used to infer the effect of specific transcription factors or microRNAs [17], which may not have been directly measured. Inference of the action of such "hidden" transcription factor or microRNA variables can be achieved by measuring the transcription factor's or microRNA's targets or associated pathways.

In the next paragraphs, we will elaborate on these concepts. We will first explain the theoretical basis underlying network models in biology and then discuss how network approaches can be employed to solve real world problems.

10.2 Networks: What Are They Really?

Networks are both an abstract mathematical concept and, due their intuitive interpretability, an excellent tool for research process communication. Very different concepts – biological and mathematical – can therefore be described in network terms. For most wet lab biologists and physicians, networks are representations of the chemical interactions seen in biochemistry textbooks (Fig. 10.2a). In this view, each node represents a specific molecule and edges account for possible physical interactions. Examples include Protein-Protein Interaction (PPI) networks, signaling pathways, transcription factor (TF)-DNA networks and metabolic networks. For each edge there is an experimental procedure that can experimentally verify the interaction and pointers to the literature can be provided as evidence. In contrast, the mathematical modeling and statistics network perspective is quite different to this "functional" biological perspective. Here a network is used to provide structure for a model. A model in this context is a representation of a biological mechanism with mathematical parameters and relationships, which enables prediction of the described phenomenon. In this model context, nodes might represent completely different concepts derived from biological, physico-chemical, geographical or chronological domains. For instance measured protein concentrations, expression of a gene, age of the patient and number of patients with a specific condition treated in the hospital can all be important to predict the outcome of a medical intervention and may all be included in network representations.

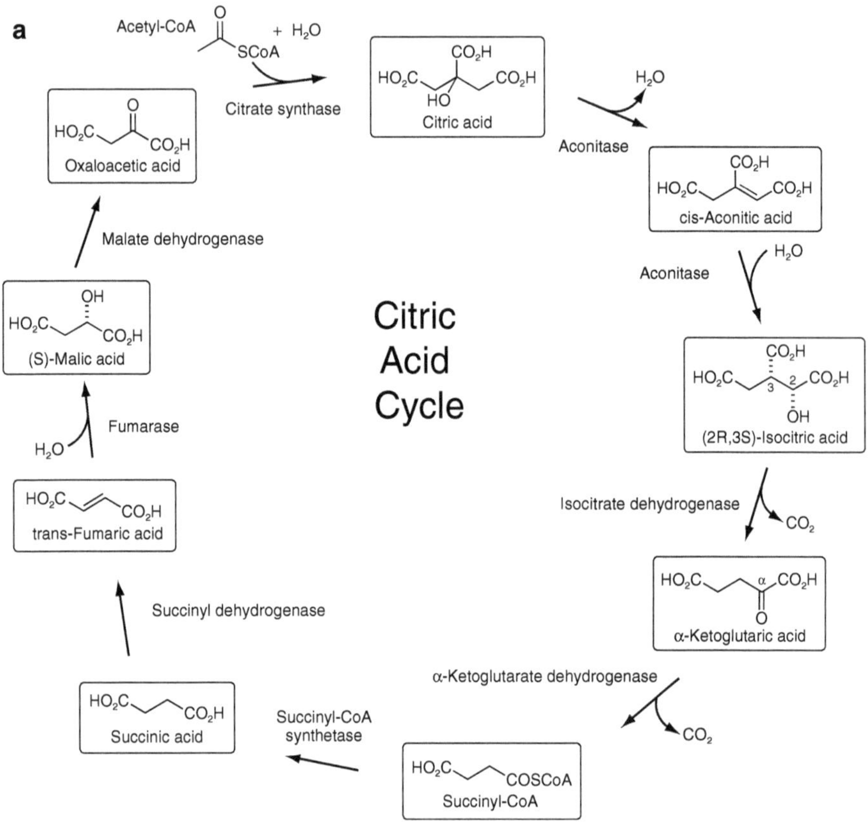

Fig. 10.2 Examples of real world and theoretical networks. **(a)** The citric acid (Krebs) cycle as classical example for a metabolic network. A map with landmarks **(b)** can be turned into a matrix by measuring the distances between each pair of landmarks **(c)**. These relationships can be visualized as a heatmap **(d)** and networks **(e-g)** Different layouts can highlight various network properties or modularity **(g)** with graphical networks that mirror the intuitive "geographical interpretation" to a greater or lesser extent

10.2.1 "Known" Biological Networks: What Do We Really Know?

With biology having arrived in the high-throughput era, diverse network realms are actively being mapped with the ultimate goal to eventually exhaustively map all regulatory interactions (for example, all functional PPIs or TF-DNA interactions). Biology textbooks with "classical" examples of metabolic networks (Fig. 10.2a) have set the expectation that eventually we will be able to construct similar maps for all existing cellular networks, and much effort has been directed toward this task. For example, several large-scale efforts have provided experimentally verified data for repositories such as the Human Protein Reference Database (HPRD, www.hprd.org). The current HPRD Release nine contains staggering numbers of proteins (30,047), PPIs (39,194) curated from 453,521 PubMed listed studies. However, in

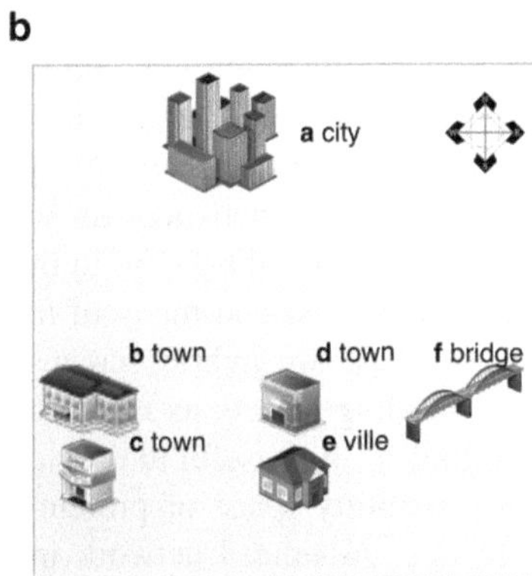

c

	a	b	c	d	e	f
a	0	184	222	177	216	231
b	184	0	45	123	128	200
c	222	45	0	129	121	203
d	177	123	129	0	46	83
e	216	128	121	46	0	83
f	231	200	203	83	83	0

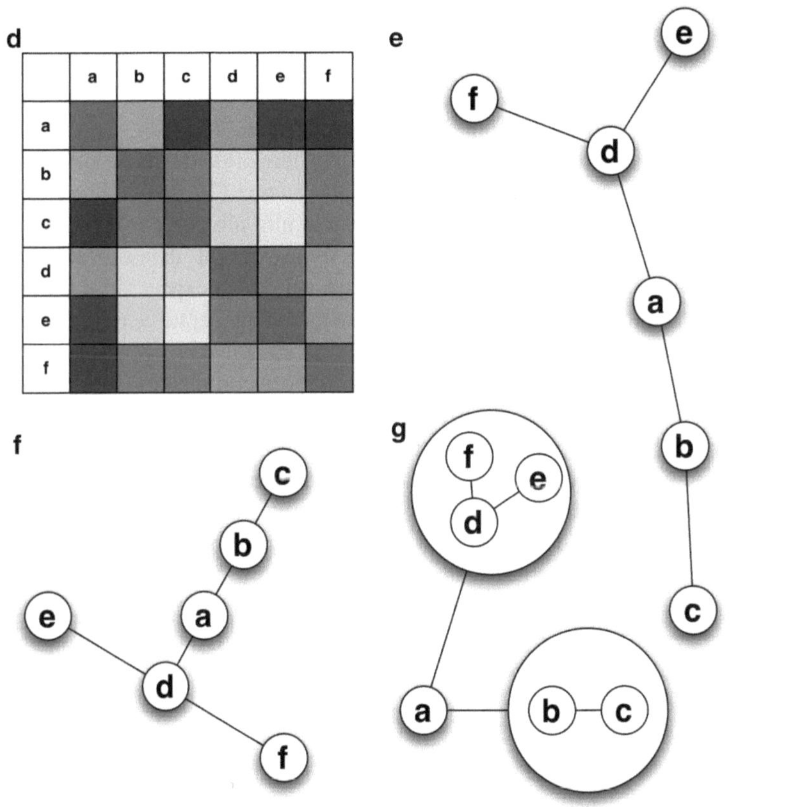

Fig. 10.2 (continued)

spite of these enormous figures we are far from a complete experimentally verified PPI network. Current estimates put the number of human PPIs to be approximately 650,000 [20]. Similarly, despite significant efforts we still know relatively little about TF-DNA networks and intracellular pathways. For example, the TRANSFAC database (www.biobase.com) contains 17,587 factors, 43,178 factor site links and 25,191 references; while the KEGG database (http://www.genome.jp/kegg/) contains 98 pathway maps, 8,494 biochemical reactions and 140,607 references. Additionally, such databases include a surprisingly high number of false positive "experimentally verified" interactions. Again, using PPI networks as example, current estimates for false positive "hits" lurking in our databases are in the range of a staggering 43–90% [21, 22] and it has been observed that many of the high-throughput datasets (for example, arising from yeast-two-hybrid screens) show very little overlap, presumably due to large sampling effects as result of small technical variation in these experiments. Nevertheless, it is useful to use data from these sources as a way of reducing the vast possibility space of protein-protein interactions to a computationally manageable size. Advanced network inference algorithms like MATISSE [23, 24] solve the problem of high false positive discovery rates as well the extreme re-discovery bias by only including nodes and edges which are supported by functional genomics data and by allowing the inference of "backnodes" in PPI networks, which represent computationally inferred proteins and edges based entirely on functional genomics data.

In many settings the most easily derivable networks are those inferred solely from data without any prior known biological structure [25, 26]. Public repositories like Gene Expression Omnibus (http://www.ncbi.nlm.nih.gov/geo) or ArrayExpress (http://www.ebi.ac.uk/arrayexpress/) make thousands of datasets accessible. If there are datasets available in these databases where we expect similar interaction structures to our experimental data we can use them to generate "external" networks. In this regard, the Dream3 challenge [27] has shown that even simple network inference algorithms can regularly outperform more complex models. Since "data-based" models are unlikely to infer the exact structure of the true network, these approaches are often combined with a coarse-grained, module based approach. Either only a "meso-scale" module structure is inferred [28] or, as in the case of the WeiGhted Correlation Network Analysis algorithm (WGCNA) [29], a full network is inferred and reduced to a lower detail network in several steps. In the foreseeable future we will have to combine our partial network knowledge with functional datasets, such as microarray gene expression compendia, both of which, when analyzed alone, are incapable of extracting truly mechanistic insights into a particular cellular phenotype. Furthermore, additional data, for instance regarding culture conditions, may also be available and can significantly increase the predictive power of network models. For practical and routine application all these heterogeneous data sources need to be bundled by one algorithm that should be able to meaningfully integrate all the relevant input data.

Furthermore, if we are interested in mechanistic modeling (where one usually has to focus on understanding localized subsets of the underlying regulatory mechanisms such as metabolic flux-, signaling pathway- and transcription factor networks in detail) we need much higher quality information, especially if we need to infer parameters that are not observed. For instance, in the stem cell context, there is now an enormous amount of mechanistic knowledge concerning core pluripotency networks and networks involved in differentiation [30]. We believe that pluripotent

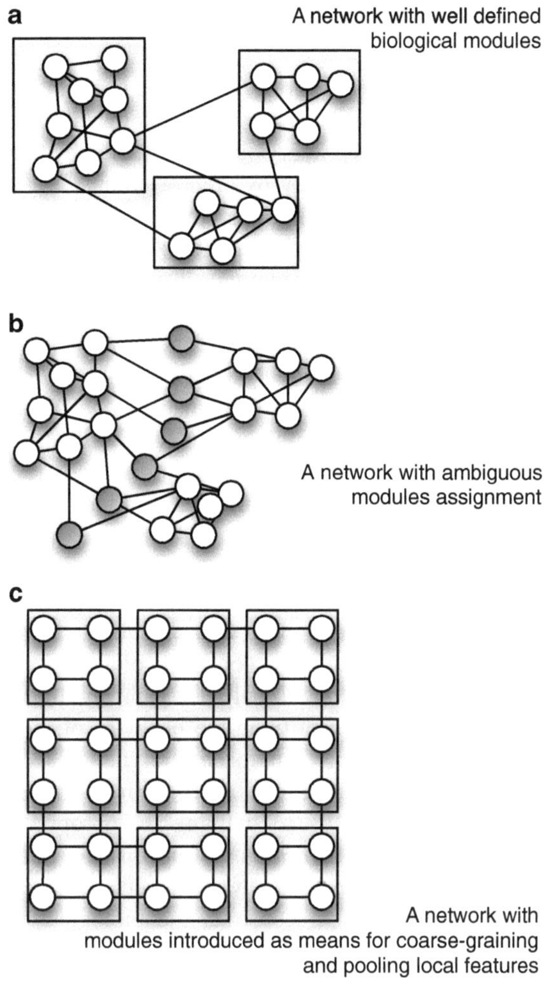

Fig. 10.3 Modules and coarse-graining. (**a**) Modules (indicated by squares) are groups of biological factors (e.g. gene or proteins, indicated by circles) that interact more strongly with each other than with those outside the module. An "ideal" modular structure as depicted in (**a**) usually implies a modular separation into non-ovelapping biological functions. Superimposition of artificial modules on real world networks is necessary for resolving issues with ambiguous module assignments for certain factors (**b**). Coarse-graining and renormalization can be useful even in the complete absence of biological modular structures of any kind as such artificial modularization allows for dimension reduction and data compression (**c**)

stem cells are now the best studied mammalian cell type, but still there are, and will be for a considerable time, large gaps in our mechanistic understanding of even the most basic cellular principles.

As take-home message of the previous paragraphs, we would like to reiterate that inferred (and even more so currently measured "real" biological) networks are inherently mechanistically unreliable. So how much of a problem is this? With false positive rates for PPIs in the range of guessing and even the largest datasets being extremely biased towards only a subset of "interesting" genes, we believe that it is reasonable to take a pragmatic approach and focus on networks that "just work for you" (that is, provide robust predictions), even if they might be only inferred and not mechanistically verifiable.

10.2.2 A Pragmatic Approach to Network Modeling

As our focus is on network performance (in terms of data interpretation, and correct hypothesis generation) rather than biological "truth", we have to invest significant effort to assess this primary outcome measure. Often it is possible to improve model performance significantly by preprocessing the networks derived in earlier steps. For example, one can remove edges that are not supported by a certain kind of evidence (see for example Fig. 10.1) or focus on a single data source such as PPI networks, thus not considering all edges from the network. Independent of the one "true" network there a certain classes of data-driven, theoretical networks that have interesting properties from a modeling perspective. Enforcing sparsity (small feature sets), using aggregate variables for a coarse grained model, restricting the available networks to a manageable subclass such as trees or completely disconnected networks might not necessarily bring us closer to a single "true" mechanistic representation but might nonetheless vastly improve the model's "real world" predictive performance and reduce computational complexity.

An alternative to the "true" mechanistic representation including all components in detail may be the restriction of the network model to mesoscopic structures. Mesoscopic modeling means that groups of network components showing similar behavior are lumped together and are substituted by a new joined variable. Prominent examples are pathways consisting of a multitude of tightly interacting proteins. If these proteins belong to only one pathway, then they can be substituted by only one new variable quantifying the behavior of the entire pathway without loss in accuracy of the model. Approaches relying on such mesoscopic module structures have the advantage that they can be derived both directly from data or from actual networks such as PPIs, by defining an appropriate distance measure and using a suitable clustering algorithm on the distance matrix (see Fig. 10.2c, d). If the features/genes are biologically organized as distinct parts, a well chosen clustering algorithm is able to reconstruct the biological partitions into groups from the data (Fig. 10.3a). Yet, crosstalk and shared membership between modules and pathways often make such a clear-cut partition unrealistic. While the microstructure (based experimentally observable data) and macroscopic structures (such as the nucleus, endoplasmatic reticulum, cell or the whole organism) possess defined biological interpretations, computational mesoscale models represent artificial modeling choices.

The association of specific genes to a mesoscopic structure such as a signaling pathway or cellular phenotype is often non-unique. In this case, soft-clustering approaches such as non-negative matrix factorization (NMF) can give rise to more realistic models. For example, there might be no compelling reason to add a certain gene to one mesoscopic module over another as these "synthetic features" may closely mirror the behavior of two distinct pathways, both of which the gene takes part in. Many different approaches can lead to equally acceptable and highly predictive data partitions. An extreme case is illustrated in Fig. 10.3c, the well-known discretization problem in physical space. For example, the grid in Fig. 10.3c might correspond to a digital image in which groups of pixels are combined to generate an image with lower resolution. In these instances there is typically no "correct" way to partition the data. Since these procedures are based on modeling decisions that rely on arbitrary choices it important to check how these choices affect the outcome. If one does not want to rely on random choice of a structure, ensemble approaches, in which many different options are tested, can be used [31, 32].

10.3 Evaluation of Network Models

Once we have generated a network model with an algorithm of our choice, the most important step is to evaluate the model performance. Network inference is commonly evaluated by comparison with known biological "gold-standard" networks, for instance by comparing with known and experimentally verified PPI or transcription factor binding sites that should have been picked up by the algorithm. However, currently, direct interactions between genes (such as PPIs or TF-DNA interactions) inferred from indirect high-content data such as gene expression measurements do not match very well with directly observed biological gold standard networks. Some of the spurious interactions arise from indirect interactions and network motifs that are inherently difficult to model from high-throughput data alone [27]. For this reason, it is much more appropriate (and rigorous) to evaluate a network model's predictive capabilities using independent datasets in a classification context. That is, by defining summary statistics that can be related to the observable outcomes (for example: is a sample pluripotent according to the teratoma assay?) Modeling choices can then be evaluated by the standard procedures of statistical learning theory.

In almost all modern network reconstruction projects the number of available samples (n) is considerably less than the number of features/variables (p) [$n \ll p$]. Network inference algorithms are designed to find patterns and may, therefore, introduce many new parameters to a model that is already haunted by this "curse of dimensionality". It is to be expected that a large set of patterns, many of which are false positives, will fit equally well to the data. If the set of "right" patterns cannot be identified *a priori*, then the set of possible patterns, which have to be checked for fit to the data, grows exponentially with the number of variables to be included in the pattern. As a result, spurious patterns in the training data (Fig. 10.1) can be picked up, resulting in unreliable predictions in independent data sets. Somewhat counter-intuitively, reducing the number of parameters to be fitted can help to ameliorate this over-fitting problem, in addition to adhering to other well established modeling practices and workflows (see [26] for details of machine learning workflows).

Statistical learning aims at improving a model fit by minimizing some definition of error. There are basically three different kinds of error:

1. Lack of fit on the training data.
2. Lack of fit on samples from the same population but not included in the training set. For example, a pluripotent sample from an experiment carried out in the same lab as the training set. This is sometimes called the "out-of-sample error".

3. The extrapolation error. How well does our model fit on data from different labs and on closely related cell types that where not included in the training population?

All common algorithms are able to improve the fit on the training data. Therefore, in cell biology the most important objective is to improve the out-of-sample error and to estimate the range of extrapolation. The out-of-sample error can be established algorithmically in the common cross-validation framework [26]. To begin with, the available dataset is split randomly into two or more independent subsets: a training set and a test set. Then, the network patterns are selected by means of standard minimization of residuals to the data using the training set only. The quality of the pattern selection is then assessed on the test set, which provides a gauge for the out-of-sample error. To start we may use networks with low "complexity" (for instance, network size), in order to guarantee good numerical stability. However, due to the low complexity, the residuals of both training and test data set will be high even if the parameter fitting has been well performed. Then, we increase the complexity of the network, again fitting on the training dataset and testing performance on the test dataset. At first, stepwise increases in the complexity will reduce monotonically the residuals on the training data. However, beyond a critical complexity (for instance, at a critical network size) the residuals on the test set will increase due to the onset of over-fitting. Hence, this "critical complexity" of the pattern identification algorithm indicates the turning point when over-fitting starts to become problematic and no increase in reliability of the patterns can be expected by increased model complexity. Thus, this critical complexity may be used to find the best compromise between model performance and over-fitting.

While there are theoretical results on the extrapolation capabilities of classical mechanistic systems and hybrid systems including some parameters learned from data [49], there is in general no automatic procedure to test the extrapolation error. This step is therefore often absent in the literature or not used to its full potential. Independent data sets [1, 24] can be used to assess a model's predictive capacity. Such additional data sets can be found and compiled from publicly available expression databases (for instance from the NCBI GEO or ArrayExpress microarray repositories) and enable assessment of the quality of the model against data observed "in the wild". To really establish the extrapolation range, one needs to construct new, focused, datasets based on both biological and mathematical understanding of the classification task [1, 24]. For example, if one tries to generate a network-based classification system for a particular stem cell type, it is highly informative to confront the classifier with biological data that is known to be difficult (such as data from differentiation time-courses, directed conversion and cell type mixing experiments). This step can sometimes provide a harsh reality check by revealing the true utility of an inferred model. From our own experience we know all too well how disappointing results from such rigorous tests can be. Yet, this validation step protects from highly unpleasant surprises when other groups cannot replicate findings with their own data, which is currently one of the most problematic issues associated with microarray studies [7–9, 34]. As Richard

Feynman famously pointed out, it is easy to fool oneself and it is extremely important to not get too enthusiastic about seemingly confirmatory results from network reconstruction algorithms that seem to fit nicely to existing biological knowledge, since such findings might arise purely by chance.

Every method has its limits and it is very important to test aggressively these boundaries. It is unfortunate that the bioinformatics literature routinely reports only positive results for practical applications, and thus presents an overly optimistic view on the real world performance of bioinformatic algorithms. As a result, in our work we specifically challenge our models until we expose the weaknesses and vulnerabilities of our classifier. For example, we have previously derived a predictive computational model for pluripotency called PluriTest [1]. To challenge this model, we prepared samples with varying mixtures of pluripotent stem cells, neural stem cells and fibroblasts as well as derived partially reprogrammed cells. These samples were not included in the "conventional" training and test datasets, but were used after model establishment and validation to identify the actual, real world performance of the PluriTest model. This "driving the model over the cliff" allows us to take precautionary measures, such as implementation of specific microarray chip quality controls. Also, a good workflow should possess a warning when the results are unreliable because of crossing the aforementioned model boundaries, not simply a true/false answer to the research question at hand.

10.4 Some Examples

In the following section we discuss some examples from our own work, specifically on applications of network-based dimension reduction methodologies to classifying stem cell populations. We want to begin this section with briefly explaining the theoretical basis for such algorithms to non-specialists and present a network interpretation of the resulting models.

An embryonic stem cell sample's gene expression profile, as assessed by microarray for example, can be considered from a mathematical standpoint as providing the coordinates of a point in a high-dimensional "expression" space in which every gene represents an axis (or dimension). Now, most genes do not behave independently, but rather are co-regulated with other genes and thus show coordinated patterns of expression. For example, the core pluripotency module consisting of POU5F1, SOX2, NANOG and other genes displays a highly correlated pattern in diverse PSC types. Thus, much of the information contained in gene expression profiles is often somewhat redundant and the dataset as a whole can be mathematically projected into fewer, more manageable dimensions without too much distortion. Algorithms by which high dimensional data is reconstructed in a lower dimensional space are known as dimensionality reduction methods, of which Principal Components Analysis (PCA) (see Fig. 10.4a) is probably the most widely known, but by no means the only, approach. For example, we have used a related method, known as non-negative matrix factorization (NMF), extensively [1, 24].

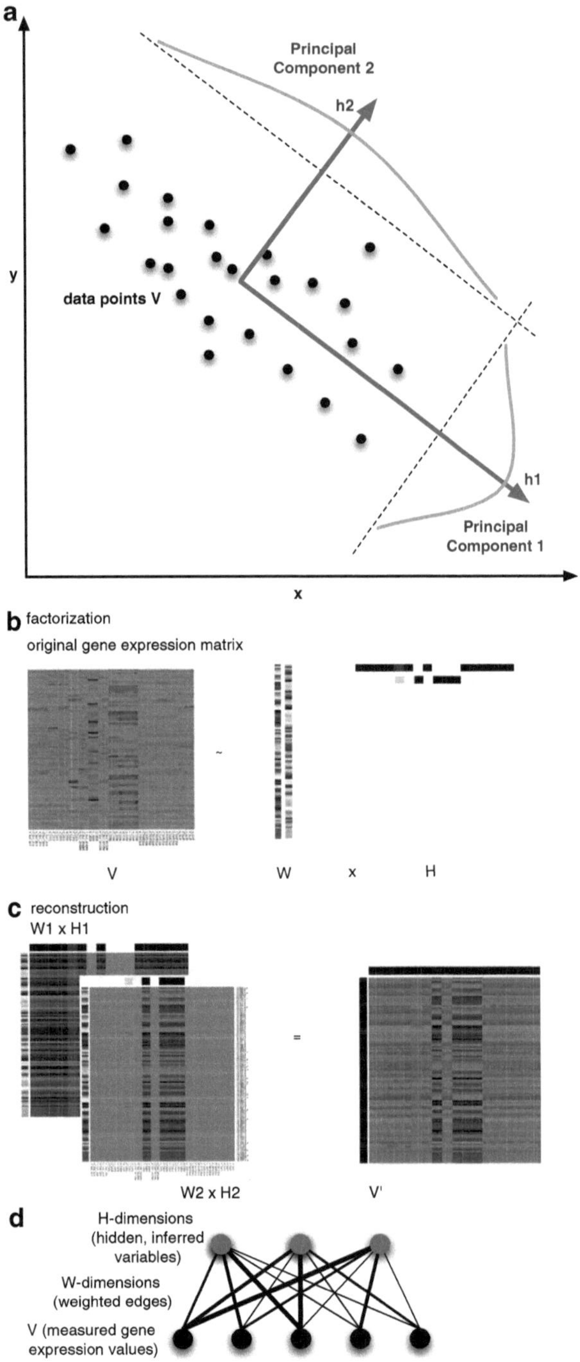

Fig. 10.4 Dimension reduction: Principles and examples. (**a**) illustrates the concepts underlying Principal Component Analysis (PCA). Data points are reduced by fitting vectors into the main components of variation of all data points. In PCA, the data dimensions are reduced in a sequential manner starting with the vector capturing largest variation. The next dimension is required to be exactly orthogonal to the previous dimension. Another, very similar method for dimension reduction is non-negative matrix factorization (NMF). In the case of NMF, a data matrix V is decomposed (**b**) into two matrices (W and H,), which, after multiplication will result in a matrix V' approximating V (**c**). The reduced dimensions in W and H do not need to be orthogonal on each other. In contrast to PCA, NMF requires the reduced dimensions to be non-negative, thus allowing for a certain degree of interpretability of the resulting base vectors. One interpretation for the NMF procedure is considering entries in the H- and V-matrix as nodes and entries in the W-matrix as weighted edges of a network

Typically, NMF is used as a sample clustering [32] or bi-clustering algorithm [34], assigning each gene to the most strongly supported cluster in order to find clusters of genes and conditions that show similar behavior. However, NMF [35] also has a network interpretation as is illustrated by the following *gedankenexperiment*. Instead of inferring a "true" network of interactions between the genes NMF infers hidden variables (groups of related genes) that can be thought of as "metagenes" (Fig. 10.4d) [32, 35]. For example, NMF could pick up the core pluripotency module from the above example as a "pluripotency metagene" and thus detect in samples (for example, cells undergoing differentiation) different levels of the pluripotency signature.

NMF metagenes are conceptually similar to the principal components in PCA and can be effectively used to model collective states of biological processes [36]. Each "real world" gene possesses a connection to each metagene (in particular, each gene can be influenced by more than one metagene), but no direct connections among metagenes (or among genes) are inferred by NMF. The weights of the gene-metagene interactions are given in a matrix known as the W-matrix (Fig. 10.4b–d). This soft-clustering approach alleviates the strict module membership problem inherent to many other algorithms. Once a W-matrix has been identified by NMF, the problem of fitting values for latent metagenes to a sample is reduced to a simple, computationally inexpensive, convex optimization problem. NMF can be used for clustering the samples "meta-expression patterns" and outperforms for class discovery a number of other, less involved algorithms in regard to biological interpretation and reliability [24, 32, 37].

Figure 10.4b shows the heatmap of experimentally measured gene expression values for pluripotent stem cells and the resulting NMF factorization. From these matrices one can reconstruct an approximation of the original data (Fig. 10.4c). Different methods can be evaluated by how well they reconstruct the data.

10.4.1 PluriNet: Beyond "Stemness"

In 2002 two seminal papers were published back-to-back in the prestigious journal Science [38, 39] attempting to identify a "quintessential" molecular signature for stem cells. Underlying these attempts was the assumption that stem cells – which have the same regenerative function in different organ systems – might possess a shared molecular machinery that, once understood, could be used to control stem cells for regenerative purposes. One of the two manuscripts coined the term "stemness" for the almost "magical" self-renewal and differentiation capabilities of stem cells.

While it is often easy and highly unfair to retrospectively criticize influential studies published at the beginning of a field, we would like to critically highlight the remaining and important legacy of these two papers and their influence on network research in the stem cell context. While neither study was the first to use high-content methods (they were significantly predated in this regard by other studies [40–42]) they represented the starting point for mainstream use of high-content nucleotide methods in stem cell biology. Both groups combined sophisticated biological experimentation across several organ systems with high content data analysis and thus can be regarded as seminal starting point of todays highly evolved stem cell systems biology field. Common to both studies was an intuition-driven attempt at network identification through co-expression. On study used presence and absence calls in combination with Venn-diagrams to find and visualize gene communities that were expressed among neural, hematopoietic and embryonic stem cells [38]. However, the devil lies in the details and, as an technical comment on both studies published the following year in Science [43] pointed out, the results of these studies did not overlap, particular when other groups tried to replicate these initial findings using other wet lab and array methodologies. So what was the problem?

Since we had previously observed a large heterogeneity among neural stem cells both *in vitro* and *in vivo*, we began to address this question by assuming that the underlying hypothesis of a general "stemness" network itself was inadequate. Thus, we began to test an alternative null hypothesis: despite apparent functional similarities, diverse stem cell types could have distinct transcriptional profiles and diverse underlying molecular regulatory networks. We supposed that there is no single shared stem cell self-renewal mechanism. To test this hypothesis we compiled a large database of gene expression microarray datasets from diverse human stem cell populations and importantly, differentiated somatic cells [24]. Assuming pluripotent stem cells share a common pervasive transcriptional program (different from somatic cells), we reasoned that an unbiased class discovery algorithm should be able to separate pluripotent stem cells from somatic cells. To achieve this we applied NMF in a consensus-clustering framework [32]. NMF is good for this purpose, since it allows for decomposition of large and heterogeneous gene expression datasets and uses co-regulated networks of genes for class discovery. Additionally, NMF can be tuned to use smaller and more localized feature sets [32, 37]. However, no matter how we tuned our algorithm, there was no single instance in which we could reliably separate embryonic and neural stem cells in one class from somatic cells in another. Following the stemness-hypothesis, such a separation should have been obvious and easy to achieve. This result was also mirrored when we used a wide range of alternative unbiased class discovery algorithms [24]. The picture that emerged was one in which human embryonic stem cells were fundamentally different from all other cell types while other stem cell types could be identified as distinct classes on what appeared to be a somatic-stem cell continuum.

Although NMF can provide a semantically or biological meaningful interpretation of its base feature sets [35, 37], we found the resulting gene sets difficult to interpret. Thus, we chose another method to elucidate experimentally derived protein-protein interaction networks, that were up-regulated in the various stem cell classes. The methodology we used was based on MATISSE [23], an algorithm that identifies so called "Jointly Active Connected Sub-graphs" (JACS) by in networks derived from gene expression datasets and PPIs. The rationale behind this choice was that PPIs are an intense subject of study for wet lab stem cell biologists. Consequently, we reasoned that identifying stem cell specific characteristics through a network-based approach could provide intuitive and biologically interpretable characteristics specific to the stem cell classes discovered with NMF.

As example we picked a network, which we called PluriNet, that was specific for human embryonic stem cells and which, surprisingly, captured a large amount of knowledge on pluripotent stem cells using unsupervised and unbiased machine learning algorithms. We discovered with independent datasets, that PluriNet had remarkable properties. It was able – after transposition into the mouse model system – to predict pluripotent features in other stem cell preparations with surprisingly high reliability. PluriNet was also able to predict that the then newly-discovered induced pluripotent stem cells were actually nearly indistinguishable from human embryonic stem cells, even when different microarray platforms were used for interrogating the respective stem cell transcriptomes. The network also was able to highlight important similarities in the transcriptomes of human MII oocytes and pluripotent stem cells.

Our results underscore a number of points, which exemplify the utility – and also the challenges – of applying network-based algorithms in exploratory studies. Firstly, networks (as with any other biological or statistical method) can be highly misleading if the underlying experimental hypothesis has not been properly tested against an appropriate null hypothesis. Secondly, abstract, data-driven methods, such as PCA [44] or NMF [35] can be extremely powerful tools for biological subsystem discovery [45], class discovery [32], and supervised classification [1] and, in our experience, outperform biologically motivated, intuitive approaches such as simple stem cell marker based assays. Thirdly, the combination of unsupervised network algorithms with unbiased, but extremely noisy, data sources such as microarray expression data and experimental PPI data can serve as "relevance filters" which meaningfully separate biological mechanisms from experimental false positives. For example, PluriNet was generated from microarray data, somewhat fuzzy stem cell class assignment and experimentally derived PPI data. On its own, none of these data sources can be used as a reliable predictive biomarker for pluripotency. However, in combination, the resulting network could predict successful reprogramming of somatic cells towards pluripotency using data from different microarray platforms and even species.

10.4.2 PluriTest: Beyond Teratomas

There is an ongoing and widespread need for wet lab experimentalists to quickly and reliably assess distinct qualities in stem cell preparations before these in vitro cultures can be used in other biological experiments, for instance modeling certain human disorders in differentiation experiments [46]. Therefore, after the publication of PluriNet in 2008 we received many requests from wet lab scientists looking to test a diverse array of stem cell preparations with PluriNet. The current "gold standard" for assessing pluripotency is still the teratoma assay: an experiment in which pluripotent stem cells are injected into immunodeficient mice and teratoma formation is assessed. This assay is costly, requires animal experimentation, and most importantly takes 2–3 months to obtain results. This led us to consider developing a more general phenotypic assay for pluripotent stem cells. For this we used a model that employs network concepts without using previously inferred networks. In order to make the procedure as flexible as possible, we decided to use a mesoscopic, purely data-driven approach which could be readily extended to additional, less well studied, cell types and even cancers. With this in mind, we found it particularly important to rely on experimental data alone and not to use potentially biased networks curated from the literature [47]. As in our previous study, NMF offered a reliable, intuitive and interpretable soft clustering framework for a data-driven network partitioning. Within this methodological framework we have implemented a bioinformatic algorithm for assaying pluripotency [1]. The algorithm is as follows: first we decompose the sample data in an unsupervised fashion with NMF and select those W-dimensions (metagenes), which best separate pluripotent stem cells from somatic cells. From these W-dimensions, we then compute, for each new sample, the expression levels of the "pluripotency metagenes", summarize them in a pluripotency score, and compare this score with those found in pluripotent stem cells.

To indicate samples that lie outside of the range of extrapolation of our classifier or which show novel features that might indicate defects, we use a one-class classification approach that is based on a consistency approach [48]. To do so, we decompose all pluripotent samples without including any somatic samples. In this case, we can ask if the gene expression profile of a new sample can be approximated with the W-dimensions from the training dataset (see Fig. 10.4c, d). If the reconstruction error passes a certain threshold, it is safe to assume that features in the tested stem cell preparation do not fit our pluripotency model and thus should be regarded as novel. This parameter (termed by us the "novelty score") is a sensitive measure of early differentiation processes in pluripotent stem cell samples and can also be used to identify and highlight localized genetic and epigenetic abnormalities in pluripotent stem cell preparations, which are otherwise indistinguishable from normal human embryonic stem cells or induced pluripotent stem cells if one only tests for conventional pluripotency markers or pluripotency associated gene signatures.

We tested this statistical framework on several independently generated datasets and found it to be stable across different microarray platforms. We have also produced an easy to use web-based implementation (www.pluritest.org), which we have called PluriTest, which should simplify the rapid testing of several hundreds of novel stem cell samples for non-bioinformaticians (see Fig. 10.5). Our future work in this area will be concerned with exploring pluripotent substates in pluripotent stem cells, projection of signatures across species boundaries and expanding the target phenotypes beyond the well understood pluripotency model system.

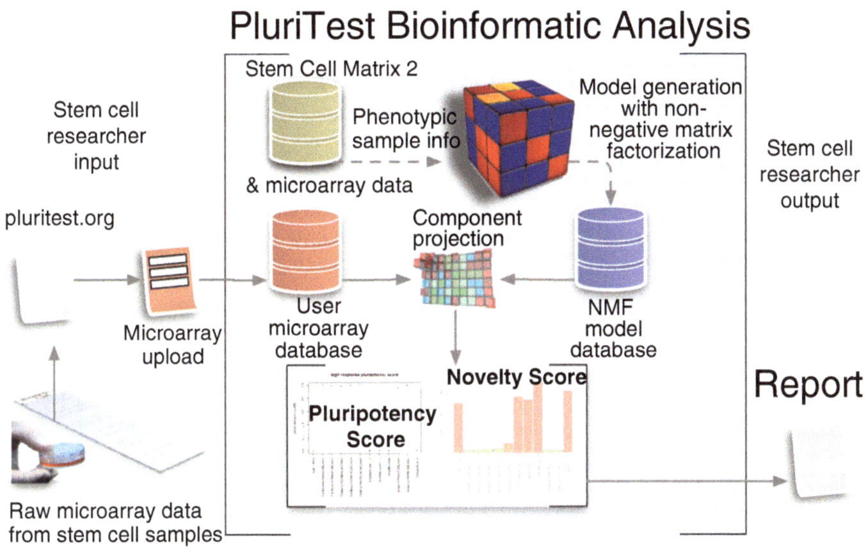

Fig. 10.5 PluriTest interaction workflow

10.5 Conclusions

Biological network modeling methodologies are extremely powerful tools for discovering biological mechanisms in stem cells. As we and others have shown, such *in silico* models, in combination with high-content datasets, can be used to reliably classify stem cells and judge differentiation potential and propensities. Current challenges in this area concern generating network models that are generalizable to more than one dataset and allow the generation of biologically meaningful predictions that can inform structured wet lab experiments.

One of the biggest challenges in this area is the establishment of a common language between the experimental biology and modeling sides of a project. This is not a one-way street: traditionally many mathematical insights have resulted from the interaction with real world-applications in physics, engineering and chemistry. In our own experience, the challenge from biology and chemistry applications has helped to generate new directions of research in mathematics (for example: hybrid models; integrating a priori known network structures into black-box machine learning algorithms for network reconstruction; and mesoscopic reconstruction of networks [33–51]). While many of these topics are out the scope of this short review, we believe that the widespread use of these approaches will effectively augment, complement and in some instances supersede wet lab experiments in the near future. As well as providing key insights that can be only be derived from analyzing complex data structures and relationships, the use of network algorithms will result in more rapid turn-around times for biological experiments. Ultimately, such assays may help facilitate a faster transition from preclinical proof-of-principle experiments to clinical applications. For this vision to become a reality, the next generation of biomedical scientists will need to be trained with expertise in mathematical modeling and wet lab experimentation. Bridging this knowledge gap and the "comfort zones" of wet lab experimentalists and theoreticians will enable the realization of today's systems biology promises.

Acknowledgements We thank Qiong Lin, Michael Lenz and Jeanne Loring for valuable discussions. FJM is supported by an Else-Kröner Fresenius Stiftung fellowship. BMS is supported by Bayer Technology Services GmbH and the Deutsche Forschungsgemeinschaft [GSC 111].

References

1. Müller F-J, Schuldt BM, Williams R, Mason D, Altun G, Papapetrou EP et al (2011) A bioinformatic assay for pluripotency in human cells. Nat Methods 8:315–317
2. Schöler HR, Hatzopoulos AK, Balling R, Suzuki N, Gruss P (1989) A family of octamer-specific proteins present during mouse embryogenesis: evidence for germline-specific expression of an Oct factor. EMBO J 8(9):2543–2550
3. Schöler HR, Balling R, Hatzopoulos AK, Suzuki N, Gruss P (1989) Octamer binding proteins confer transcriptional activity in early mouse embryogenesis. EMBO J 8(9):2551–2557
4. Takahashi K, Yamanaka S (2006) Induction of pluripotent stem cells from mouse embryonic and adult fibroblast cultures by defined factors. Cell 126(4):663–676
5. Kim JB, Greber B, Araúzo-Bravo MJ, Meyer J, Park KI, Zaehres H et al (2009) Direct reprogramming of human neural stem cells by OCT4. Nature 461(7264):649–653
6. Andrews PW, Fenderson B, Hakomori S (1987) Human embryonal carcinoma cells and their differentiation in culture. Int J Androl 10(1):95–104
7. Leek JT, Scharpf RB, Bravo HC, Simcha D, Langmead B, Johnson WE et al (2010) Tackling the widespread and critical impact of batch effects in high-throughput data. Nat Rev Genet 11(10):733–739
8. Chin MH, Mason MJ, Xie W, Volinia S, Singer M, Peterson C et al (2009) Induced pluripotent stem cells and embryonic stem cells are distinguished by gene expression signatures. Cell Stem Cell 5(1):111–123
9. Newman AM, Cooper JB (2010) Lab-specific gene expression signatures in pluripotent stem cells. Cell Stem Cell 7(2):258–262
10. Christodoulou C, Longmire TA, Shen SS, Bourdon A, Sommer CA, Gadue P et al (2011) Mouse ES and iPS cells can form similar definitive endoderm despite differences in imprinted genes. J Clin Invest 121(6):2313–2325
11. Bock C, Kiskinis E, Verstappen G, Gu H, Boulting G, Smith ZD et al (2011) Reference maps of human ES and iPS cell variation enable high-throughput characterization of pluripotent cell lines. Cell 144(3):439–452
12. Leek JT, Storey JD (2007) Capturing heterogeneity in gene expression studies by surrogate variable analysis. PLoS Genet 3(9):1724–1735
13. Boulting GL, Kiskinis E, Croft GF, Amoroso MW, Oakley DH, Wainger BJ et al (2011) A functionally characterized test set of human induced pluripotent stem cells. Nat Biotechnol 29(3):279–286
14. Morizane A, Doi D, Kikuchi T, Nishimura K, Takahashi J (2011) Small-molecule inhibitors of bone morphogenic protein and activin/nodal signals promote highly efficient neural induction from human pluripotent stem cells. J Neurosci Res 89(2):117–126

15. Chambers SM, Fasano CA, Papapetrou EP, Tomishima M, Sadelain M, Studer L (2009) Highly efficient neural conversion of human ES and iPS cells by dual inhibition of SMAD signaling. Nat Biotechnol 27(3):275–280
16. Thomson M, Liu SJ, Zou L-N, Smith Z, Meissner A, Ramanathan S (2011) Pluripotency factors in embryonic stem cells regulate differentiation into germ layers. Cell 45(6):875–889
17. Kim H, Lee G, Ganat Y, Papapetrou EP, Lipchina I, Socci ND et al (2011) miR-371-3 expression predicts neural differentiation propensity in human pluripotent stem cells. Cell Stem Cell 8(6):695–706
18. Kalman R (1959) On the general theory of control systems. IRE Trans Autom Control 4:481–492
19. Liu Y-Y, Slotine J-J, Barabási A-L (2011) Controllability of complex networks. Nature 473:167–173
20. Stumpf MPH, Thorne T, de Silva E, Stewart R, An HJ, Lappe M et al (2008) Estimating the size of the human interactome. Proc Natl Acad Sci USA 105(19):6959–6964
21. Deane CM, Salwiński Ł, Xenarios I, Eisenberg D (2002) Protein interactions: two methods for assessment of the reliability of high throughput observations. Mol Cell Proteomics 1(5):349–356
22. Huang H, Bader JS (2009) Precision and recall estimates for two-hybrid screens. Bioinformatics 25(3):372–378
23. Ulitsky I, Shamir R (2007) Identification of functional modules using network topology and high-throughput data. BMC Syst Biol 1:8
24. Müller F-J, Laurent LC, Kostka D, Ulitsky I, Williams R, Lu C et al (2008) Regulatory networks define phenotypic classes of human stem cell lines. Nature 455(7211):401–405
25. Markowetz F, Troyanskaya OG (2007) Computational identification of cellular networks and pathways. Mol Biosyst 3(7):478–482
26. Hastie T, Tibshirani R, Friedman J (2009) The elements of statistical learning: data mining, inference, and prediction, 2nd edn. Springer, New York
27. Marbach D, Prill RJ, Schaffter T, Mattiussi C, Floreano D, Stolovitzky G (2010) Revealing strengths and weaknesses of methods for gene network inference. Proc Natl Acad Sci USA 107(14):6286–6291
28. Wong DJ, Liu H, Ridky TW, Cassarino D, Segal E, Chang HY (2008) Module map of stem cell genes guides creation of epithelial cancer stem cells. Cell Stem Cell 2(4):333–344
29. Langfelder P, Horvath S (2008) WGCNA: an R package for weighted correlation network analysis. BMC Bioinformatics 9:559
30. MacArthur BD, Ma'ayan A, Lemischka IR (2009) Systems biology of stem cell fate and cellular reprogramming. Nat Rev Mol Cell Biol 10(10):672–681
31. Joshi A, De Smet R, Marchal K, Van de Peer Y, Michoel T (2009) Module networks revisited: computational assessment and prioritization of model predictions. Bioinformatics 25(4):7
32. Brunet J-P, Tamayo P, Golub TR, Mesirov JP (2004) Metagenes and molecular pattern discovery using matrix factorization. Proc Natl Acad Sci USA 101(12):4164–4169
33. Fiedler B, Schuppert A (2008) Local identification of hybrid models with tree structure. IMA J Appl Math 73:449–476
34. Guenther MG, Frampton GM, Soldner F, Hockemeyer D, Mitalipova M, Jaenisch R et al (2010) Chromatin structure and gene expression programs of human embryonic and induced pluripotent stem cells. Cell Stem Cell 7(2):249–257
35. Pascual-Montano A, Carmona-Saez P, Chagoyen M, Tirado F, Carazo JM, Pascual-Marqui RD (2006) bioNMF: a versatile tool for non-negative matrix factorization in biology. BMC Bioinformatics 7:366
36. Kim PM, Tidor B (2003) Subsystem identification through dimensionality reduction of large-scale gene expression data. Genome Res 13(7):1706–1718
37. Gao Y, Church G (2005) Improving molecular cancer class discovery through sparse non-negative matrix factorization. Bioinformatics 21(21):3970–3975

38. Ramalho-Santos M, Yoon S, Matsuzaki Y, Mulligan RC, Melton DA (2002) "Stemness": transcriptional profiling of embryonic and adult stem cells. Science 298(5593):597–600
39. Ivanova NB, Dimos JT, Schaniel C, Hackney JA, Moore KA, Lemischka IR (2002) A stem cell molecular signature. Science 298(5593):601–604
40. Phillips RL, Ernst RE, Brunk B, Ivanova N, Mahan MA, Deanehan JK et al (2000) The genetic program of hematopoietic stem cells. Science 288(5471):1635–1640
41. Loring JF, Porter JG, Seilhammer J, Kaser MR, Wesselschmidt R (2001) A gene expression profile of embryonic stem cells and embryonic stem cell-derived neurons. Restor Neurol Neurosci 18(2-3):81–88
42. Terskikh AV, Easterday MC, Li L, Hood L, Kornblum HI, Geschwind DH et al (2001) From hematopoiesis to neuropoiesis: evidence of overlapping genetic programs. Proc Natl Acad Sci USA 98(14):7934–7939
43. Fortunel NO (2003) Comment on " 'stemness': transcriptional profiling of embryonic and adult stem cells" and "a stem cell molecular signature" [I]. Science 302(5644):393b–393
44. Pearson K (1901) On lines and planes of closest fit to systems of points in space. Phil Mag Ser 62(11):559–572
45. Lee DD, Seung HS (1999) Learning the parts of objects by non-negative matrix factorization. Nature 401(6755):788–791
46. Müller F-J, Goldmann J, Löser P, Loring JF (2010) A call to standardize teratoma assays used to define human pluripotent cell lines. Cell Stem Cell 6(5):412–414
47. Som A, Harder C, Greber B, Siatkowski M, Paudel Y, Warsow G et al (2010) The PluriNetWork: an electronic representation of the network underlying pluripotency in mouse, and its applications. PLoS One 5(12):e15165
48. Tax D, Muller K-R (2004) A consistency-based model selection for one-class classification 2004. In: ICPR 2004. Proceedings of the 17th international conference on pattern recognition, vol 3, Cambridge, UK, CA, pp 363–366
49. Schuppert A (1999) Extrapolability of structured hybrid models: a key to the optimization of complex processes. In: Bernold F, Konrad G, Juergen S (eds) Proceedings of the international conference on differential equations, Berlin, Germany, 1–7 August 1999. World Scientific Publishing, Singapore, pp 1135–1151
50. Schuppert A (2011) Efficient reengineering of Meso-scale topologies for functional networks in biomedical applications. J Math Ind 1:6
51. Schuppert A, Burghaus R, Von Törne C, Schwers S, Stropp U, Kallabis H (2006) Method for identifying predictive biomarkers from patient data. Patent WO/2007/07/9875

Index

A. Ma'ayan and Ben D. MacArthur (eds.), *New Frontiers of Network Analysis in Systems Biology*, DOI 10.1007/978-94-007-4330-4,

MIX
Papier aus verantwortungsvollen Quellen
Paper from responsible sources
FSC® C105338

If you have any concerns about our products,
you can contact us on
ProductSafety@springernature.com

In case Publisher is established outside the EU,
the EU authorized representative is:
Springer Nature Customer Service Center GmbH
Europaplatz 3, 69115 Heidelberg, Germany

Printed by Libri Plureos GmbH
in Hamburg, Germany